The
Sea Island
Mathematical
Manual

伏戲倉精初造王業畫卦結繩以理海內

THE SEA ISLAND MATHEMATICAL MANUAL: SURVEYING AND MATHEMATICS IN ANCIENT CHINA

FRANK J. SWETZ

THE PENNSYLVANIA STATE UNIVERSITY PRESS
UNIVERSITY PARK, PENNSYLVANIA

Library of Congress Cataloging-in-Publication Data

Swetz, Frank.
The sea island mathematical manual : surveying and mathematics in ancient China / Frank J. Swetz.

p. cm.
An annotated translation and discussion of Liu Hui's Haidao Suanjing (Sea island mathematical manual), written in 263 A.D.
Includes bibliographical references and index.
ISBN 0-271-00795-8 (cloth). — ISBN 0-271-00799-0 (paper)
1. Surveying — China — History. 2. Mathematics, Chinese — History.
I. Liu, Hui, 3rd/4th cent. Hai tao suan ching. 1992. II. Title.
TA527. C6S94 1992
526'. 0931 — dc20 91-13002
CIP

It is the policy of the Pennsylvania State University Press to use acid-free paper for the first printing of all clothbound books. Publications on uncoated stock satisfy the minimum requirements of American National Standard for Information Sciences — Permanence of Paper for Printed Library Materials, ANSI Z39.48–1984.

FRONTISPIECE: Tomb relief from the Han dynasty depicts Fu Xi (伏羲), the first of the legendary Three Sovereigns, with his consort Nu Wa (女娲). The divine couple are credited with bringing order to the Universe. They are usually represented with dragon tails intertwined and holding a set-square and a drawing compass, symbols of order and organization.

觸類而長之則雖幽遐詭伏靡所不入

In comprehending by analogy, problems
are always solvable though
they may be very troublesome
and obscure.

—Liu Hui (A.D. 263)

CONTENTS

PREFACE

Almost all ancient societies practiced the art of land survey, but few extant records describe their measuring techniques and specific mathematical procedures. In Pharaonic Egypt, a system of royal taxation was levied on landholdings. Because the Nile flooded its banks annually, priest-surveyors periodically had to reconfirm and reestablish boundary markers to ensure accurate tax assessments. Although little is known about their techniques, they were recognized in the ancient world as competent surveyors from whom other peoples acquired the same skill. Legend credits Thales of Miletus (625?–547? B.C.) with employing Egyptian "shadow-reckoning" techniques to determine the height of the pyramids. He brought his knowledge of Egyptian surveying and geometry back to Greece. The Greeks transformed the empirical geometry of the Egyptians into a formal deductive system based on abstraction but did little to develop the rudiments of the surveying they had learned. Application of geometry to the general problems of land mensuration was considered intellectually degrading. Despite this stigma, several Greek mathematicians were noted for their ability to determine the height of mountains by means of indirect measurement. Among these were Dicaearchus of Messina (fl. ca. 320 B.C.), a disciple of Aristotle; Eratosthenes of Alexandria (fl. ca. 230 B.C.), the first prominent geographer of antiquity; and Heron of Alexandria, who flourished during the first century A.D. Heron was a practical surveyor and engineer who compiled and examined traditional Greek and Egyptian surveying methods to produce a text of his own on the subject, *Treatise on the Dioptra*. Heron's work, in turn, served as a primary source of theory for later Roman surveyors, the *agrimensores* or *gromatici*, and established traditions that would be followed in European surveying through the Middle Ages. Although the art of surveying was important for the political and military well-being of the Roman Empire, and although the Romans produced impressive physical structures, the *agrimensores* also did little to advance the theoretical basis of their discipline. Existing Roman documents appear to focus more on the legal aspects of surveying work rather than on methodology or the mathematical principles involved.

Thus, early Western surveying literature reveals little about the practical uses of geometry and mathematics in the tasks of land mensuration. But recent research into the content and origins of early Chinese mathematics is revealing that there were strong traditions and interest in the methodologies and applications of land survey. It is from Chinese sources that a clearer picture of how mathematics and geometry were adapted to the needs of surveying emerge.

This monograph examines and discusses the concern of the ancient Chinese with the development of land surveying techniques and the mathematics those techniques involve. A single work—the *Haidao Suanjing*, or *Sea Island Mathematical Manual* (A.D. 263)—is examined in depth, and its contents are analyzed for their mathematical implications. The *Haidao Suanjing* marks a high point of theoretical and mathematical sophistication in Chinese surveying theory and sets the standards for much of East Asian surveying activity for the next one thousand years after its appearance. Its problem situations and solutions provide insights into the Chinese applications of right triangle theory and reveal a mathematical systemization previously unrecognized in Chinese works. This monograph complements and extends the findings of a previous study on the Chinese understanding and use of the right triangle: *Was Pythagoras Chinese? An Examination of Right Triangle Theory in Ancient China* (1977), by the author and his associate T. I. Kao. Together, these two monographs testify to early Chinese priority and superiority in the mathematical utilization of right triangle principles.

In any research study of this scope, many people are involved in various ways. I would like to acknowledge and thank those who assisted me in developing and completing this book. Ang Tian Se of the Department of Chinese Studies at the University of Malaya initially collaborated with me in researching and writing "A Chinese Mathematical Classic of the Third Century: The *Sea Island Mathematical Manual* of Liu Hui," which appeared in the journal *Historia Mathematica* (13 [1986]: 99–177) and serves as a basis for the present study. In particular, Ang did the translation of the *Sea Island* problems. Unfortunately, he could not continue on the project with me. Otto Bekken of Agder College, Kristiansand, Norway, reviewed an earlier draft of the manuscript and made several important suggestions. His observations prompted some revision and helped produce a better historical study. I thank Academic Press, publisher of *Historia Mathematica*, for allowing me to use and reproduce material from my previous published article in this book. I would also like to thank Howard Sachs, Associate Dean for Research at The Pennsylvania State University at Harrisburg, and the College's Research Committee, for supporting and encouraging my work; Janice Russ of the Research Office at The Pennsylvania State University at Harrisburg for her patient and expert typing of the manuscript; and Wang Jiang-chao for his assistance with the Chinese calligraphy.

1

PERSPECTIVES

The Importance of Surveying to the Chinese Empire

In 1973, excavations of a Han dynasty tomb at Mawangdui in the region of Changsha in southwest China uncovered three ancient maps of a marquisate.[1] The maps dated from the early Western Han period (206 B.C.–A.D. 24). All three were painted on silk and, despite their age, had features that are still clearly discernible. Perhaps the most impressive item in this collection was a topographical rendering of what today can be recognized as the Xiaoshui River basin. This map was drawn to a scale of approximately 1:180,000 and clearly depicts mountain ranges, the courses of rivers, population centers, and basic ground contour configurations. When compared with a modern-day map of the same region, the topological features recorded more than two thousand years ago are remarkably accurate. The ancient surveyors who made this map clearly possessed high

1. See Cao Wanru[co], "Maps 2,000 Years Ago and Ancient Cartographical Rules," in Institute of the History of Natural Sciences, Chinese Academy of Science, *Ancient China's Technology and Science* (Beijing: Foreign Language Press, 1983), pp. 250–57.

levels of technical and mathematical skill.

The existence and use of maps are frequently mentioned in such pre-Qin dynasty (211–207 B.C.) Chinese classics as the *Zhou Li*[a]* (Rites of the Zhou Dynasty), *Zhan Guo Ce*[b] (Records of the Warring States), and *Guan Zi*[c] (The Book of Master Guan), which in offering the following advice clarifies the importance of early Chinese maps:

> Before a military campaign the commander must read the maps carefully for terrains so steep or waterlogged as to hamper or damage carts, for possible valleys or passes through impassable mountain ridges, and for dense forests of underbrush where enemy ambushes threaten.
>
> He must learn all the necessary details such as the length of routes, sizes and possible strongholds of cities, and even causes of prosperity or decline of towns. He must memorize all these geographical features and set great store by them. Only thus can he succeed in maneuvers and raids, in unfolding his every strategic or tactical step logically, and in exploiting the topographical advantages to the full. Maps are used most often for military purposes.[2]

It is interesting to note that the second of the excavated marquisate maps is a military map depicting troop locations within the described territory. Maps were even included in military classics, such as *Sun Zi Bing Fa*[d] (Master Sun's Art of War) (fifth century B.C.) and *Sun Bin Bing Fa*[e] (Sun Bin's Art of War) (fourth century B.C.). In their construction, military maps demanded a high degree of reliability with regard to distances indicated, orientations, and the locations of mountain ranges, river courses, roads, and population centers. The tasks of achieving this reliability fell to surveyors, whose measurements and calculations decided these factors.

Pei Xiu[f] (A.D. 223–271) is considered the father of Chinese cartography. In 267 he was appointed minister of works by the first emperor of the unified Jin dynasty (A.D. 265–316). Pei collected and studied all existing maps of the empire and concluded that good mapmaking involved adherence to six principles:

1. The use of an appropriate scale in drawing.
2. The employment of a rectangular grid system.

---

* Superscript letters refer to the equivalent Chinese characters, found in the Glossary at the end of the present book. Characters are supplied for certain proper names, titles, and important terms.

2. As quoted in ibid., pp. 250–51.

3. Accurate measurement for distances between major landmarks. In instances where such landmarks were at different elevations, their location had to be accurately projected onto a plane. This projection process required further accuracies.
4. The determination of elevations.
5. The measurement of right and acute angles.
6. The measurement of curves and straight lines.

Each of these principles depended on an understanding and use of the mathematical properties of right triangles. This knowledge formed the basis of Chinese surveying and cartographic practice.

While the needs of military and civil mapmaking were a major focus for Chinese surveying activities, surveying had developed primarily to satisfy the imperial needs of engineering and construction. From ancient times China has been dependent on water control and conservancy. Early Chinese society was river-based and utilized the fertile land bordering such great rivers as the Yangtze and Hwang Ho (Yellow River) for agricultural purposes. But Chinese rivers are flood-prone and unpredictable—they had to be harnessed and controlled, and systems of dikes, canals, and irrigation channels were built for this purpose. In the process of developing and perfecting such structures and systems, geometry, land mensuration, and surveying techniques and theories were devised and recorded. The new and powerful knowledge was soon absorbed into tradition and myth. Wall reliefs on the Wu Liang tomb shrines (ca. A.D. 140) found in Shandong Province depict Fu Xi[g], a mythical ruler and folk hero, holding two drawing compasses in his hands and his consort Nu Wa[h] holding a set-square.[3] The L-shaped set-square, similar to that used by carpenters, was a principal surveying instrument. The legendary emperor-engineer Yu[i] the Great is also often depicted holding a set-square, an instrument he supposedly used to tame the waters of China, in his hands:

> Emperor Yu quells the floods, he deepens rivers and streams, observes the shape of mountains and valleys, surveys the high and low places, relieves the greatest calamities and saves the people from danger. He leads the floods east into the sea and insures no flooding or drowning. This is made possible because of the *Gougu* right triangle theorem.[4]

---

3. Depicted in Li Yan[bs] and Du Shiran[cp], *Chinese Mathematics: A Concise History*, trans. John N. Crossley and Anthony Lun (Oxford: Clarendon Press, 1987), p. 2.

4. Commentary of Zhao Shuang on the *Zhoubi Suanjing* as quoted in ibid., p. 29.

FIG. 1

Thus land surveying in China has an extensive history and from the earliest times was an important official activity closely associated with the well-being of the empire.

Surveying Tools of the Ancient Chinese

From surviving pictures, artifacts, and descriptions, we know about the principal surveying instruments employed by the ancient Chinese—namely, sighting or reference poles, *biao*[j] (see Figure 1a); the set-square or gnomon, *ju*[k]; the plumb line, *xian*[l]; the water level, *zhun*[m] (Figure 1b); and ropes and

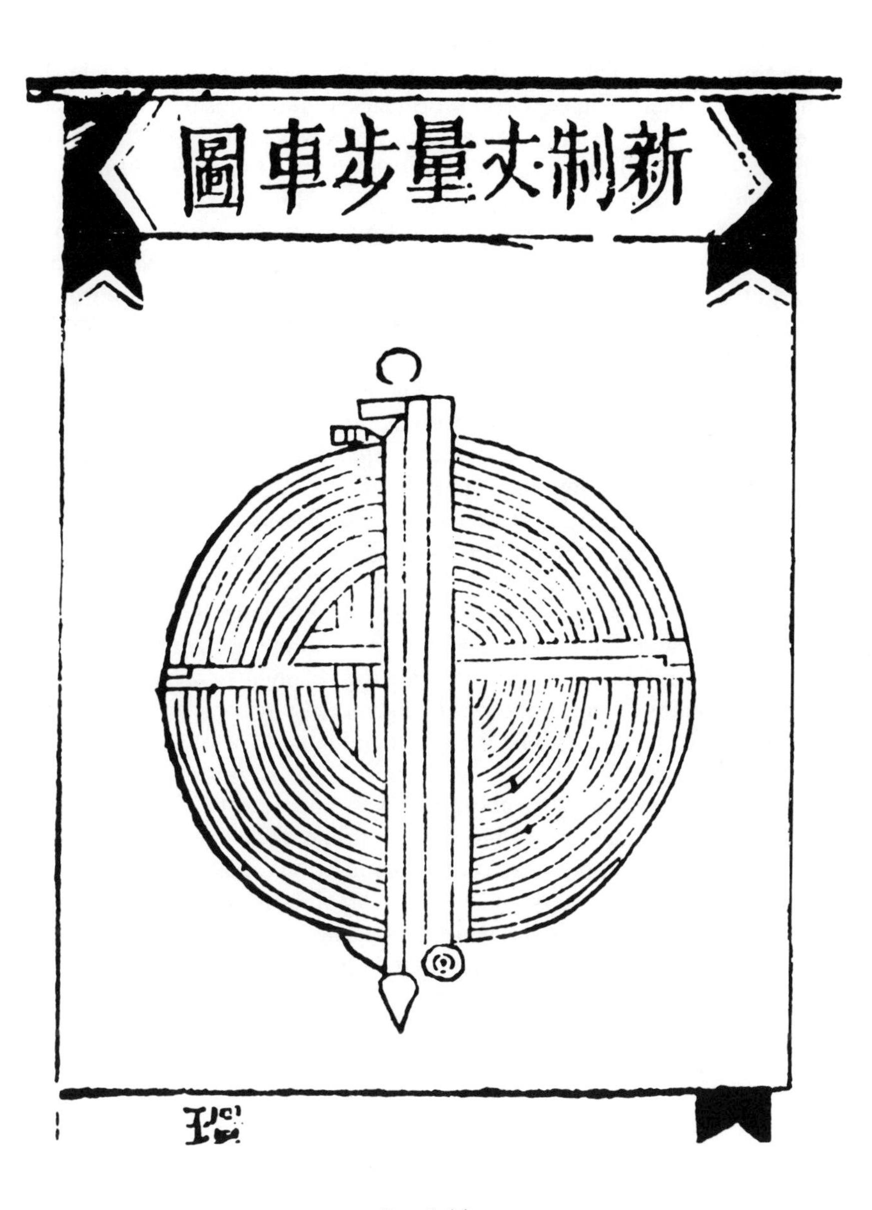

FIG. 1 (c)

cords, which were later replaced by measuring tapes, *bu che*[n] (Figure 1c). The drawing compass, *gui*[o], is also associated with surveyors but was not a field instrument used as a measuring tool. The use of these tools was elementary and straightforward. Sightings along three poles established collinearity. Linear distances could be marked and measured by stretched ropes and cords. A vertical alignment of poles was ensured by the use of a plumb line, and sightings were secured through the use of sighting tubes, *wang tong*[p], or sighting boards, *ce shi pai*[q]. Nonvertical sightings were obtained with the assistance of the set-square. This instrument probably evolved from the use of a vertical staff and a horizontal shadow template placed at its base. The determination of shadow lengths was important in timekeeping and astronomical observation.[5] When the template was joined to the staff, the L-shaped set-square was formed. A surveyor's version of the set-square was portable, had legs about two feet long, could be hand-held, and was easily transported. The use and versatility of the set-square, or *ju*, is described in the *Zhoubi Suanjing*[r] (The Arithmetical Classic of the Gnomon and the Circular Paths of Heaven) (ca. 100 B.C.–A.D. 100), the oldest extant Chinese mathematical classic, by means of a rather fanciful conversation between Zhou Gong[s], a duke of the Zhou dynasty (ca. 1030–221 B.C.) and the Grand Prefect Shang Gao[t]. The duke inquires, "May I ask how to use the set square?" To which Shang Gao replies, "Align the set square with the plumb line to determine the horizontal, lay the set square down to measure height, reverse the set square to measure depth, lay the set square down to determine distance. By revolving the set square about its vertex a circle can be formed, combining two set squares forms a square" (see Figure 2).[6]

The use of the set-square, combined with appropriate observations and measurements, made it possible to determine distances to inaccessible points. In such situations, necessary computations were based on simple principles of proportionality involving similar right triangles, but eventually a series of prescribed formulas were devised for surveying. As we shall see, these formulas were mathematically sound and ingeniously conceived.

---

5. See the discussions of shadow observations in Joseph Needham, *Science and Civilization in China*, vol. 3 (Cambridge University Press, 1959); Frank J. Swetz, "Trigonometry Comes Out of the Shadows," in *Proceedings of the Kristiansand Workshop on the History of Mathematics, Kristiansand, Norway, August, 1989* (forthcoming).

6. As translated in Li and Du, *Chinese Mathematics*, p. 31.

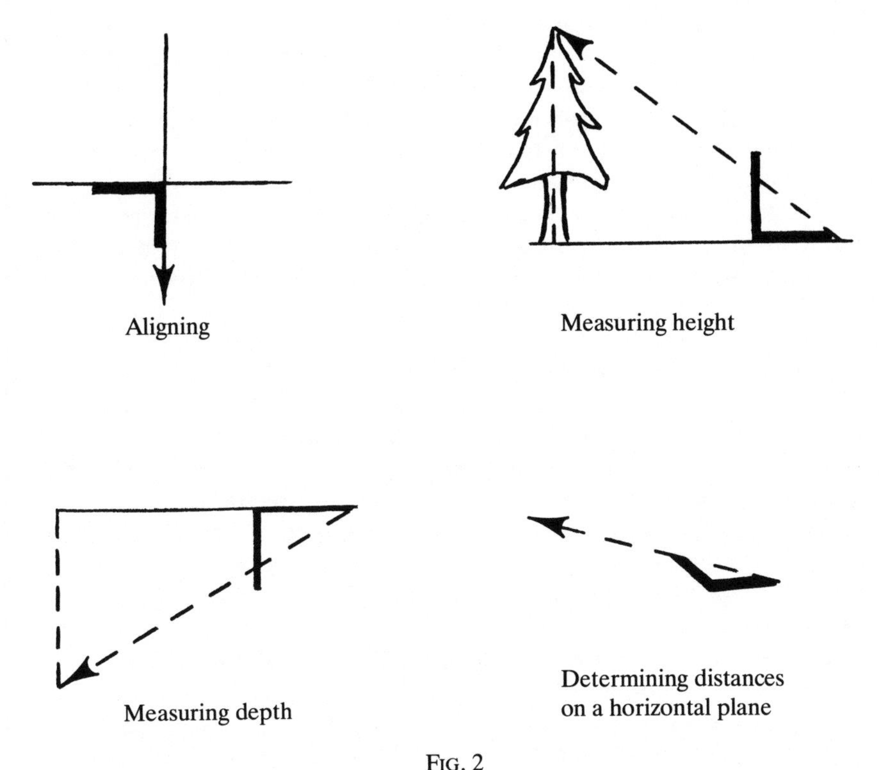

FIG. 2

The *Haidao Suanjing*[u] and Its Tradition in Chinese Surveying Literature

All the earliest known Chinese mathematical classics contain material either on or directly pertaining to surveying. Mathematics texts were written for official use and reference. Their contents reflected the mathematical priorities of the empire, and land mensuration and surveying were such priorities.

Although the *Zhoubi Suanjing* was probably compiled into its present form as a text in the period 100 B.C.–A.D. 100, its contents reflect material known and used at a much earlier period of Chinese history. In translation, the book's title refers to "perimeter" (*zhou*)[v] and "gnomon," or shadow casting staff or pointer (*bi*)[w]. Thus Joseph Needham, an eminent sinologist and historian of science, named it *The Arithmetical Classic of the Gnomon and Circular Paths of Heaven*, a title that has since been popularly accepted.

While the *Zhoubi* is generally considered a text on astronomy, its discussion of right triangle theory and applications and the use of measuring instruments and techniques also pertain to the practices of land surveying. It should be noted that astronomers were called "surveyors of heavens," and in early times the terms for astronomer and land surveyor were exactly the same: *chou jen*[x].

Next in historical order of appearance is the *Jiuzhang Suanshu*[y] (The Nine Chapters on the Mathematical Art), a general compilation that spans the accomplishments of the Zhou, Qin, and Han eras (ca. eleventh century B.C.–A.D. 220) and was probably formed as a text in the same period as the *Zhoubi*. A bureaucratic handbook, the *Jiuzhang* contains 246 specific problems divided into nine specialized chapters.[7] Three chapters are especially relevant to the needs of surveyors: chapter 1, "Field Measurements"; chapter 5, "Construction Consultations"; and chapter 9, "*Gougu*[z]" (right triangle), where eight of its twenty-four problems—problems 17–24—are directly concerned with surveying situations. In A.D. 263 the scholar-official Liu Hui[aa] wrote a commentary on the *Jiuzhang* in which he supplied theoretical verification for the recorded solution procedures and expanded the text with his own contributions. Liu believed chapter 9 inadequate in its coverage of measurements to inaccessible locations and its involvement with a method of calculation called *chong cha*[ab], or "double difference."

> ...in researching the *Nine Calculations* (Nine Chapters) there is a term "double differences,"... for finding heights or surveying great depths while knowing the distances, one must use "double differences."... As well as doing the commentary I researched the meaning of the ancients and reformulated "double differences" as an appendix to the *Gougu* chapter.[8]

Liu extended this chapter by adding nine problems of his own, each of which involved inaccessible distances and used the *chong cha* technique to obtain a solution. His contribution marked the high point of early Chinese mathematical surveying theory. During the following Northern and Southern dynasty period, the noted mathematician Zu Congzhi[ac] (A.D. 429–500) wrote a commentary on this extended version of the *Jiuzhang*. A few years later the scholar Zhang Qiujian[ad] (fl. + 468) wrote *Zhang Qiujian*

7. Discussion of the *Jiuzhang Suanshu* given in Ho Peng Yoke, *Li, Qi, and Shu: An Introduction to Science and Civilization in China* (Hong Kong: Hong Kong University Press, 1985), pp. 64–67; Frank Swetz, "The Amazing Chiu Chang Suan Shu," *Mathematics Teacher* 65 (1972): 425–30.

8. Translated in Li and Du, *Chinese Mathematics*, p. 75.

Suanjing[ae] (Mathematical Manual of Zhang Qiujian), in which he considered surveying problems similar to those conceived by Liu. At the beginning of the Tang dynasty (A.D. 618–906), the additional nine problems involving double differences were separated from the body of the *Jiuzhang* and made into an independent mathematical work, the *Haidao Suanjing* (Sea Island Mathematical Manual), the title of which was derived from the context of the first problem in the collection, which deals with determining distances involving a remote sea island. See Figure 3.

When the Tang government instituted a department of mathematics at its Royal Academy in A.D. 656, the curriculum of study was set by the contents of a specific collection of works, the *Suanjing Shi Shu*[af] (Ten Mathematical Manuals). One of these ten manuals was the *Haidao Suanjing*. The *Suanjing Shi Shu* was edited by Li Chunfeng[ag], a celebrated astronomer and mathematician. Li wrote a commentary on the *Haidao* entry, preparing it for study by the civil service candidates of the academy. While most required texts were prescribed for a year of analysis and study, the *Haidao* demanded three years of attention—an indication of its importance. By the eighth century the Tang curriculum had also been adopted for use in the territories of Korea and Japan. Thus, knowledge of the *Haidao* and its methods became known over a wide area of Asia.

During the Song dynasty (A.D. 960–1278), two editions of the *Haidao Suanjing* were published in 1084 and 1213 respectively, but these works later became lost. However, the tradition, methods, and problems of the *Haidao* lived on in the works of various mathematicians of this period. In 1247, Qin Jiushao[ah] published the *Shushu Jiuzhang*[ai] (Mathematical Treatise in Nine Sections).[9] Three of its nine chapters concerned either surveying or surveying applications: chapter 4 was entitled "Surveying"; chapter 7, "Architecture"; and chapter 8, "Military Matters." Chapter 4 contained nine problems specifically devoted to survey work, while chapters 7 and 8 concerned applications of surveying principles—for example, "Military Matters" considered the use of surveying techniques for observing an enemy from a distance.[10] Several of Qin's problem situations were taken directly from the *Haidao* collection. Although he did not seem to be secure in his knowledge of surveying mathematics, Qin's work is the first major systematic consideration of the subject since the initial appearance of the *Haidao Suanjing* some one thousand years earlier.

---

9. A detailed study of this work is given in Ulrich Libbrecht, *Chinese Mathematics in the Thirteenth Century: The Shu-Shu Chiu-Chang of Ch'in Chiu-Shao* (Cambridge, Mass.: MIT Press, 1973).

10. Ibid., p. 147.

FIG. 3 A woodblock print illustrating the sea island situation. Given in the *Tu Shu Ji Cheng* (1726), an encyclopedia.

Yang Hui[aj] of the Southern Song dynasty (1127–1280) was a diligent mathematician who sought mathematical justifications for many of the solution techniques employed by his predecessors. In several of his works he discussed the *Haidao* problems and their *chong cha* solution methods. For example, in his *Xu Gu Zhai Qui Suanfa*[ak] (Continuation of Ancient Methods for Elucidating the Strange [Properties of Numbers]) (1275), Yang presented and analyzed four problems concerning computation of inaccessible heights; one was the first problem in the *Haidao* collection.[11] It is interesting to note that in the same year Yang also published a separate work devoted to the needs of surveyors, *Tian Mu Bi Lei Cheng Chu Jie Fa*[al] (Practical Rules of Arithmetic for Surveying).

Historically, this was a period of splendid mathematical achievement in China, but still the importance of surveying theory persisted. The famous mathematician Zhu Shijie[am](1280–1303) of the Yuan dynasty (1271–1368) considered eight surveying problems in his *Siyuan Yujian*[an] (Precious Mirror of the Four Elements) (1303),[12] five of which were taken from the *Haidao*. Although Zhu was primarily interested in demonstrating the power of the *si yuan shu*[ao] algebraic solution methods, he singled these problems out as being important.[13]

Although the *Haidao*'s problems were often presented and reviewed in such piecemeal manner, the complete collection of problems as an independent mathematical entity became rare. Under Ming rule, however, the whole collection of problems was recorded, albeit haphazardly, in the *Yongle Da Dian*[ap] (The Great Encyclopedia of the Yongle Reign Period) (1403–1407).

Toward the end of the Ming dynasty, Western influence began to penetrate China. In 1582 the Italian Jesuit Matteo Ricci (1552–1610) arrived in Macao as a missionary, and a year later he entered mainland China and made his way to Beijing, where he used his scientific and mathematical knowledge to assist the Chinese with calendrical reform. Ricci assumed the scholarly name Li Madou[aq] and worked with Chinese colleagues translating and compiling European works on mathematics and astronomy into the Chinese language. In collaboration with the scholar Xu Guangqi[ar] (1562–1633), he published *Ce Liang Fa Yi*[as] (Essentials of

11. See Lam Lay-Yong, *A Critical Study of the Yang Hui Suan Fa* (Singapore: University Press, 1977), p. 345.

12. A study of this work was undertaken by Jock Hoe, *Les Systèmes d'Équations Polynômes dans le Siyuan Jujian (1303)*, Mémoires de l'Institut des Hautes Études Chinoises, vol. 6 (Paris: Collège de France, 1977).

13. Zhu generalized the "technique of the celestial element" used to solve equations in one unknown to the solution of equations in four unknowns through the "technique of four elements." See the more complete discussion in Li and Du, *Chinese Mathematics*, pp. 135–48.

Surveying [Trigonometry]) (1607–1608). This book introduced contemporary European survey and land measurement methodology to the Celestial Empire. But Ricci also realized his limitations and asked other Jesuits, who were better versed in the sciences associated with astronomy and calendrical reckoning, to visit China. Several complied and immediately occupied themselves with the tasks of calendar reform.[14]

Xu Guanqi was also skilled in the traditional methods of Chinese surveying, having devised plans for river irrigation systems in 1603. This new European knowledge fascinated him, and he set about comparing the mathematical techniques based on Euclidean concepts with those employed in contemporary China. Xu published his results as an appendix to the *Essentials of Surveying*, which appeared in 1608 under the title *Ce Liang I Tung*[aw] (Similarities and Differences [Between Chinese and European] Surveying Techniques). After Ricci's death, Xu worked with his new Jesuit colleagues in 1629–1633 to produce a monumental compendium of scientific knowledge, and he presented it to the emperor, whose name it bore. The *Chong Zhen Li Shu*[at] (Chong Zhen Reign Treatise on [Astronomy and] Calendrical Science) contained 137 chapters, each of which examined a specific scientific concept. Among its entries were several contributions on surveying, including *Da Ce*[au] (Complete Surveying), written by the Swiss Jesuit Johannes Terrentius (1567–1630), and *Ce Liang Quan Yi*[av] (Complete Theory of Surveying), by the Italian Jesuit Giacomo Rho.

Although the Ming dynasty fell under Manchu invasion from the north, official interest in Western mathematics persisted. The Manchus established the Qing dynasty (1644–1911) and adopted Chinese ways, customs, and institutions. In 1687 two French Jesuits, Jean-François Gerbillon (1654–1707) and Joachim Bouvet (1656–1730), arrived in China and were asked to live in the capital for the "purpose of serving the Court."[15] Gerbillon and Bouvet became tutors of the emperor, Kang Xi[ax], who was attracted to the study of mathematics:

> The Emperor ordered them from then [1689] on to take turns each day in the Yang Xin Hall, to lecture in Manchu [the official court language] on Western science such as surveying, etc. to His Celestial Majesty. Whenever his Majesty was free, he concentrated on learning. He loved the various disciplines: surveying, mensuration, calculating, astronomy, geometry, and logical argument.[16]

---

14. For example, Niccolo Longobardi arrived in China in 1597; Johann Terrenz Schreck in 1621; Johann Adam Schall von Bell in 1622 and Giacomo Rho in 1624.

15. Li and Du, *Chinese Mathematics*, p. 217.

16. Passages from the *Zheng Jiao Feng Bao*[cq] (Correct [Catholic] Religion Receiving Praise), in ibid.

The Manchu lecture notes compiled by these missionaries included several works pertaining to surveying, particularly *Ce Liang Gao Yuan Yi Qi Yong Fa*[ay] (Manual on the Theodolite). Under royal command this Western knowledge was compiled into reference collections. Mathematician Mei Juecheng[az] (1681–1763) directed a ten-year effort that resulted in the appearance of *Shu Li Jing Yun*[ba] (Collected Basic Principles of Mathematics) (1721), a compendium that included the mathematics of surveying. With this influx of Western knowledge much of traditional Chinese mathematical theory, including the techniques of the *Haidao*, was disregarded and forgotten.

Despite this new mathematical climate, some scholarly interest and pride in China's traditional accomplishments remained. Attempts were made to find and reconstruct lost classics. The scholar Dai Zhen[bb] (1724–1777) compiled the *Si Ku Quan Shu*[bc] (Complete Library of the Four Branches of Literature), in which he collected the *Haidao* problems from the *Great Encyclopedia* and reconstituted them as a text. This text was reproduced and used by Kong Jihan[bd] in his compilation *Suanjing Shi Shu* (Ten Mathematical Manuals) (1773), later appeared as an appendix to *Jiuzhang Suanshu* (1776) by Qu Zhenfa[be], and finally appeared in the Wu Ying Dian[bf] (Hall of Military Heros)[17] Palace edition (ca. 1794). Thus, in fifteen hundred years the *Haidao Suanjing* had traversed a complete cycle from appearing first as an appendix to the *Nine Chapters*, to existing as an independent mathematical/surveying thesis, to being fragmented and dispersed, and then finally appearing once again as an appendix to the *Nine Chapters* (see Figure 4).

Liu Hui, His Work, and His Times

By the beginning of the third century A.D., the Han dynasty was in a state of societal disarray. China was experiencing civil and intellectual strife and dissension. The centralized systems of taxation and corvée labor and military conscription had both broken down. Buddhism and Taoism were beginning to establish inroads into China, eroding traditional Confucian beliefs. Schools and libraries were destroyed. The Confucian civil service examination system, the mainstay of the imperial bureaucracy, was seriously weakened. In a historical experience reminiscent of that undergone by the Roman Empire in the West, central authority collapsed. Three generals

17. The location of one of the first Western printing presses in China.

seized power and in A.D. 211 divided the empire among themselves. General Cao Cao[bg] established his rule in the northwest territory of the last Han stronghold; in the south, Sun Quan[bh] declared himself ruler of the state of Wu with a capital in Wuchang and in Szechwan; and Liu Bei[bi] established the Shu Han dynasty. Thus, the Three Kingdom period (A.D. 220–280) came into being. In Chinese literature this time is depicted as a fanciful and romantic era, but in reality it was a time of tumult and intrigue. When Cao Cao died in A.D. 220, power was assumed by his son, Cao Pei[bj], who convinced the existing Han emperor and nominal ruler of the region to abdicate in his favor. Cao Pei assumed the title of emperor and designated the region under his rule the kingdom of Wei, with a capital at Loyang. Of the Three Kingdoms—Wei, Wu, and Shu Han—only Wei claimed direct descent from the Han. War broke out among the rival states, with Wei pressing the initiative. Gradually the other kingdoms fell to the military prowess of Wei: in 263 Shu Han was defeated and annexed, and in 280 Wu yielded to the armies of Wei.

Little is known about the life of Liu Hui except that he was an official in the kingdom of Wei during this period,[18] but from existing evidence and tradition it can be deduced that Liu Hui was a respected mathematician of some skill. In his commentary on the *Jiuzhang Suanshu*, Liu both enriched and extended its contents, providing mathematical justification for the solution procedures.[19] In many of these verifications he used models and diagrams, but these were lost, probably before the beginning of the Tang dynasty, and have since become a subject of conjecture, some of which will be discussed more closely in Chapter 3.

Several of Liu's mathematical accomplishments are worthy of note. In order to approximate a value for "pi," Liu employed a method of "circle division": using a circle with a radius of one foot he systematically inscribed regular polygons of computable area. Finally advancing to an n-gon of 192 sides, he estimated pi to have a value of 3.141024. Liu Hui realized that the process of circle division he used was theoretically infinite in nature but approached a limit in actuality. He noted, "The finer it cuts, the smaller the loss; cut after cut until no more cuts, then it coincides with the circle."[20] Liu extended this concept of limits and indivisibles by using dissection techniques to investigate and verify existing formulas for computing volumes and areas. In particular, his work on the volume of a sphere utilized theoretical methods that would later come to be known as Cavalieri's

18. See entry on Liu Hui by Ho Peng Yoke, *Dictionary of Scientific Biography*, vol. 8 (New York: Scribners, 1970–80), pp. 418–24.

19. See "The Contributions of Liu Hui," in Li and Du, *Chinese Mathematics*, pp. 65–80.

20. As quoted in ibid., p. 68.

Principle in honor of the Italian mathematician Bonaventura Cavalieri (1598–1647), who clarified and perfected the technique.[21] Liu Hui was the first Chinese mathematician to fully appreciate and use a concept of limits in his work.

No doubt Liu possessed great mathematical ability, but this ability probably would not have been realized had he not been employed in Wei at this time. The emperor of Wei sought to restore and preserve China's intellectual traditions and institutions. He ordered the copying and reconstruction of old classics and the compilation of an encyclopedia of existing knowledge. The Confucian classics were reedited, and the civil service examination system was reorganized and reinstituted in the years 237–239. Certainly for the engineering and military requirements of Wei, a reliable version of the *Jiuzhang* was needed. The kingdom's eventual domination of its neighbors has been attributed primarily to three factors:

1. Astute use of military-agricultural colonies
2. Enlargement and improvement of internal irrigation systems
3. A military strategy of prolonged attrition against enemies[22]

Each of these factors is dependent on a use of mathematical knowledge consistent with the *Jiuzhang*'s contents. Further, surveying methodologies as devised and discussed by Liu in his extended work on right triangle theory would have certainly been of interest to Wei's military commanders. Deng Ai[bk], a Wei general at that time, was noted for his knowledge and use of surveying mathematics because he always "estimated the heights and distances, measuring by finger breadths before drawing a plan of the place and fixing the position of his camp."[23] It should also be noted that the famous cartographer Pei Xiu flourished during this period. Mathematics and surveying were certainly important skills in the kingdom of Wei.

---

21. For Chinese work on the volume of a sphere, see T. Kiang, "An Old Chinese Way of Finding the Volume of a Sphere," *Mathematical Gazette* 56 (1972): 88–91; Donald B. Wagner, "Liu Hui and Tse Keng-chih on the Volume of a Sphere," *Chinese Science* 3 (1978): 59–79. Chinese involvement with "Cavalieri's Principle" is discussed in Lam Lay-Yong and Shen Kangsheng, "The Chinese Concept of Cavalieri's Principle and Its Application," *Historia Mathematica* 12 (1985): 219–28.

22. L. Carrington Goodrich, *A Short History of the Chinese People* (New York: Harper & Row, 1959), p. 58.

23. Needham, *Science and Civilization in China*, p. 572.

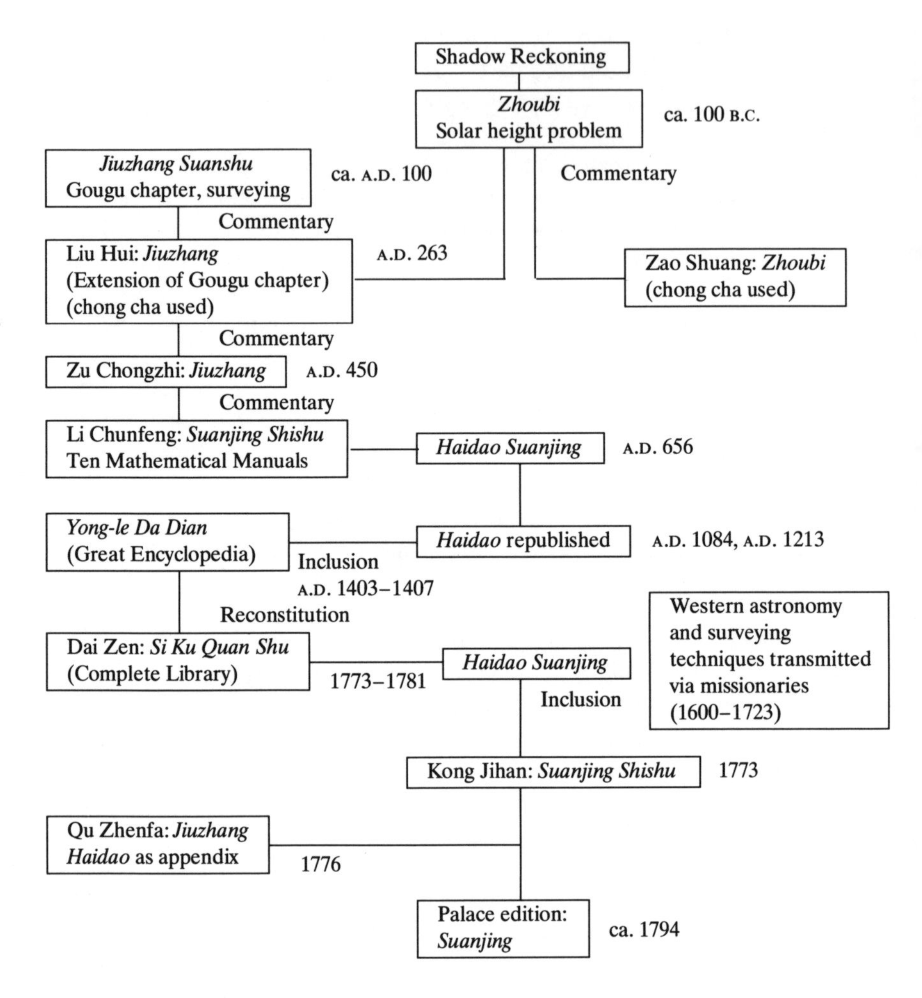

FIG. 4 Lineage and chronology of events and influences in the evolution of *chong cha* and the transmission of *Haidao Suanjing* traditional mathematical surveying theory.

2

THE *HAIDAO SUANJING*

Research on the *Haidao*

During the Qing dynasty (1644–1911) the Chinese government expelled most of the foreign missionaries who were working in its scientific bureaus, instituted a "closed-door policy" with regard to the outside world, and curtailed the flow of scientific and mathematical information into China.[1] Scholars were encouraged to examine and revise ancient indigenous

1. As foreign missionaries gained influence in China, particularly through the popularization of Western science among the educated classes, the Imperial Court feared losing its influence and respect. In 1704 the Pope decreed that Chinese Catholics should abstain from ancestor worship. This was a challenge to the traditions and customs of China and to the established order. In response, Emperor Kang Xi ordered the papal legate banned from Beijing and restricted to Macao. A conspiracy was fomented to replace the emperor with someone more sympathetic to Catholic interests, but failed, and all foreign missionaries were banned to Macao and forbidden to travel freely in mainland China. The Chinese people themselves were forbidden to travel abroad and, if abroad, were forbidden to return. This "closed-door policy" remained in effect until the Opium Wars (1840).

mathematical texts. Caught up in this chauvinistic movement, several eminent mathematicians commented on the *Haidao*'s problems and mathematical techniques. Li Huang[bl] (d. 1811) wrote the *Haidao Suanjing Xi Cao Tu Shuo*[bm] (Detailed Diagrammatic Explanation of the Sea Island Mathematical Manual), which appeared in 1819. Shen Qinpei[bn] (fl. 1836) produced a work entitled *Chong Cha Tu Shuo*[bo] (Diagrammatic Explanation of the "Double Differences") (1820). Both authors used Western Euclidean concepts to verify the solution techniques. In 1879 Li Liu[bp] wrote the *Haidao Suanjing Wei Bi*[bq] (Notes on the Sea Island Mathematical Manual), in which he employed the later traditional solution methods of *tian yuan*[br] (celestial element) to solve the sea island problems.[2] Interest in the *Haidao* subsided for a while but was resurrected by Li Yan[bs] in 1926 when he published a paper on the "double difference" method.[3] Since that time the *Haidao* has been an active subject for study among Chinese mathematical historians. The best existing version of *Haidao Suanjing* can be found in Qian Baocong's[bt] 1963 edition of the *Ten Mathematical Classics*.

In the West, the language barrier has caused interest in the subject to grow much more slowly. Yoshio Mikami's *Development of Mathematics in China and Japan* (1912) devoted a three-page chapter to a discussion of the *Haidao* in which he published an English-language translation of the first three problems. The French sinologist-mathematician L. van Hée published a translation of the same three problems in his native language in 1920,[4] but in 1932 he published a complete translation of all nine problems.[5] Van Hée's discussion of the *Sea Island Manual*'s contents was brief and contained little on the relevant history or mathematics. In a 1973 entry for the *Dictionary of Scientific Biography* on Liu Hui, Ho Peng Yoke[bu] discussed the *Haidao Suanjing* and provided English-language translations of problems 1, 4, and 7. In the Russian language a complete study of the *Haidao* was undertaken by E. I. Berezkina, who in 1974 published a translation of the problems together with her verifications of their solutions.[6] Later, in a 1980 work, Berezkina published a detailed discussion of problems 1, 3, 4, and 8.[7]

---

2. This method is described in Guo Shuchun[cr], "The Numerical Solution of Higher Equations and the *Tianyuan* Method," in *Ancient China's Technology and Science*, pp. 111–23.

3. Li Yan, "Chong Cha Shu Yuanliu Ji Qi Xin Zhu"[cs] (The Origins of the Method of Double Difference with a New Commentary), *Xue Yi*[ct] 7 (1926): 1–15.

4. L. van Hée, "Le Hai-Tao Souan-King de Lieou," *Toung Pao* 20 (1920): 51–60.

5. L. van Hée, "Le Classique de l'Île Maritime: Ouvrage Chinois du IIIe Siècle," in *Quellen und Studien zur Geschichte der Mathematik* 2 (1932): 255–58.

6. E. I. Berezkina, "Dva Teksta Lyu Khueya Po Geometrii," *Istoriko-Matematicheskie Issledovaniya* 19 (1974): 231–48.

7. E. I. Berezkina, *Matematika Drevnovo Kitaya* (Moscow: Nauka, 1980).

The first complete English-language translation of and commentary on the *Haidao* problems was accomplished by Ang Tian Se[bv] and Frank Swetz in 1986.[8] That translation is employed below.

The Problems of the *Haidao Suanjing*

The following translation of the *Haidao Suanjing* problems is based on Qian Baocong's edition of *Suanjing Shi Shu*. It is a free translation in the sense that a complete retention and rendering of archaic technical terms has been avoided. This was done to improve readability and understanding for the general reader and to eliminate the extensive annotation that would have been necessary had these terms been retained. The translation of content, however, remains historically and mathematically correct. Li Chunfeng's commentary is omitted because it does not aid in understanding the rationale of Liu Hui's solution methods. The format follows that of traditional Chinese mathematical works: a problem is presented, the solution is given, and a very mechanical solution procedure is offered. In early China all calculations were carried out using a set of computing rods and a counting board.[9] The directions for obtaining solutions to the *Haidao* problems assume the use of rods and a counting board. As used in some problems, *fa*[bw] and *shi*[bx] are technical terms indicating positions on the counting board and indicate the divisor and the dividend, respectively. Numbers have been assigned to the problems to indicate the sequential ordering of presentation and for referencing. The metrology employed in the series of problems is:

li[by]	= 1,800 *chi*[bz]
zhang[ca]	= 10 *chi*
bu[cb]	= 6 *chi*
chi	= 10 *cun*[cc]

8. Ang Tian Se and Frank J. Swetz, "A Chinese Mathematical Classic of the Third Century: *The Sea Island Mathematical Manual* of Liu Hui," *Historia Mathematica* 13 (1986): 99–117.

9. Computing rods, made of either bamboo or ivory, were laid out in vertical or horizontal configurations. These configurations represented numbers, negative and positive. In manipulating their rods, the Chinese used a positional decimal orientation that was preserved and reinforced by the use of a counting board partitioned into a matrix of square cells. Using the rods and board, Chinese mathematicians developed impressive algorithmic procedures that were later transferred to the abacus. For more detailed discussions, see Mei Rongzhao,[cu] "The Decimal Place: Value Numeration and the Rod and Bead Arithmetics, " in *Ancient China's Technology and Science*, pp. 57–65; and Ang Tian Se, "Chinese Computation with the Counting-Rods," *Kertas-Kertas Pengajian Tionghoa* 1 (1977): 97–109.

PROBLEMS:

[1] Now for [the purpose of] looking at a sea island, erect two poles of the same height, 3 *zhang* [on the ground], the distance between the front and rear [pole] being a thousand *bu*. Assume that the rear pole is aligned with the front pole. Move away 123 *bu* from the front pole and observe the peak of the island from ground level; it is seen that the tip of the front pole coincides with the peak. Move backward 127 *bu* from the rear pole and observe the peak of the island from ground level again; the tip of the back pole also coincides with the peak. What is the height of the island and how far is it from the pole?

[Answer:] The height of the island is 4 *li* 55 *bu*. It is 102 *li* 150 *bu* from the pole.

[Method:] Multiply the distance between poles by the height of the pole, giving the *shi*. Take the difference in distance from the points of observations as the *fa* to divide [the *shi*], and add what is thus obtained to the height of the pole. The result is the height of the island.

To find the distance of the island from the front pole, multiply [the distance of the] backward movement from the front pole by the distance between the poles, giving the *shi*. Take the difference in distance at the points of observation as the *fa* to divide the *shi*. The result is the distance of the island from the pole in *li*.

[2] Now for [the purpose of] looking at a pine tree of unknown height growing on a hill, erect two poles of the same height 2 *zhang* [on the ground], the distance between the front and the rear [pole] being 50 *bu*. Assume that the rear pole is aligned with the front pole. Move backward 7 *bu* 4 *chi* from the front pole and observe the top of the pine tree from ground level; it is observed that the tip of the pole coincides with the top of the pine tree. Sight again at the foot of the pine tree; its base is seen to be at a point 2 *chi* 8 *cun* from the tip of the pole. Once more, move backward 8 *bu* 5 *chi* from the rear pole and observe the top of the pine tree from ground level; it is seen that the tip of the pole also coincides with the top of

the pine tree. What is the height of the pine tree and how far is the hill from the pole?

[Answer:] The height of the pine is 12 *zhang* 2 *chi* 8 *cun*. The hill is 1 *li* 28 4/7 *bu* from the pole.

[Method:] Multiply the distance between poles by the observation entry on the pole, giving the *shi*. Take the difference in distance at the points of observation as the *fa* to divide [the *shi*] and add to it the observation measurement along the pole. The result is the height of the pine tree.

To find the distance of the hill from the pole, place the distance between poles [on the counting board] and multiply it by the distance of the rear movement from the front pole, giving the *shi*. Take the difference in distance from the points of observation as the *fa* to divide [the *shi*]. The result is the distance of the hill from the pole.

[3] Now, looking southward at a square [walled] city of unknown size, erect two poles 6 *zhang* apart in the east-west direction such that they are standing at eye level and are joined by a string. Assume that the eastern pole is aligned with the southeastern and northeastern corners of the city. Move northward 5 *bu* from the eastern pole and sight on the northwestern corner of the city; the line of observation intersects the string at a point 2 *zhang* 2 *chi* 6 1/2 *cun* from its eastern end. Once again, move backward northward 13 *bu* 2 *chi* from the pole and sight on the northwestern corner of the city; the corner coincides with the western pole. What is the [length of the] side of the square city, and how far is the city from the pole?

[Answer:] The side of the squared city measures 3 *li* 43 3/4 *bu*. The city lies 4 *li* 45 *bu* from the pole.

[Method:] Multiply the final distance from the pole by the observation measurement obtained on the string and divide [the product] by the distance between poles. What is thus obtained is the shadow difference. Subtracting the initial distance from the pole [from the shadow difference], the remainder is the *fa*. Place the final distance from the pole [on the counting board] and subtract the initial distance from the pole.

The remainder is multiplied by the observation measurement along the string, giving the *shi*. Dividing the *shi* by the *fa* yields the [length of the] side of the squared city.

To find the distance [of the city] from the pole, place the final distance from the pole [on the counting board] and subtract the shadow difference. The remainder is multiplied by the initial distance from the pole, giving the *shi*. Dividing the *shi* by the *fa* yields the distance of the city from the pole.

[4] Now, for [the purpose of] looking into a deep valley, set up a carpenter's square of height 6 *chi* on the lip of the valley. Sight on the bottom of the valley from the tip of the carpenter's square; the bottom is seen at a point 9 *chi* 1 *cun* along the base of the carpenter's square. Set up another similar carpenter's square above [the first]; the distance between the bases of the [two] squares is 3 *zhang*. Sight on the bottom of the valley from the tip of the upper square; the bottom is now seen at a point 8 *chi* 5 *cun* along the base of the upper square. How deep is the valley?

[Answer:] 41 *zhang* 9 *chi*.

[Method:] Place the distance between the carpenter's squares [on the counting board] and multiply it by the upper base, giving the *shi*. Subtract the [measurement obtained along the] upper base from the [measurement obtained along the] lower base and take the remainder as the *fa*. Perform the division, and subtract the height of the carpenter's square from it [i.e., the quotient]. The result is the depth of the valley.

[5] Now, for [the purpose of] observing a building on the level ground from a mountain, erect a carpenter's square of height 6 *chi* on the mountain. Sight on the foot of the building from the tip of the square at a downward angle; the foot is seen at a point 1 *zhang* 2 *chi* along the base of this square. Set up again another similar carpenter's square above [the first]; the distance between the bases of [upper and lower] squares is 3 *zhang*. Sight on the foot of the building from the tip of the upper square in a downward manner; the foot is now seen at a point 1 *zhang* 1 *chi*

4 *cun* along the upper base. Erect another small pole at the observation entry [on the upper base]. Sight once again from the tip of the upper square in a downward manner; the pointed gable of the building is observed at a point 8 *cun* along the erected pole. What is the height of the building?

[Answer:] 8 *zhang*.

[Method:] Subtract from one another the [measurements taken along the] upper and lower bases; take the remainder as the *fa*. Place the distance between the carpenter's squares [on the counting board], multiply it by the [measurement taken along the] lower base, and divide the product by the height of the square. The result obtained is multiplied by the observation measurement obtained on the small [vertical] pole, giving the *shi*. Dividing the *shi* by the *fa* yields the height of the building.

[6] Now, for [the purpose of] looking toward the southeast at the mouth of a river, erect two poles in the north-south direction; the distance between them is 9 *zhang*, and they are joined along the ground by a string. Face the west and move away 6 *zhang* from the north pole to observe the southern bank of the mouth of the river at ground level; the bank is seen at 4 *zhang* 2 *cun* from the north end of the string. Sight on the northern bank [from the same position]; the bank is seen [along the string] 1 *zhang* 2 *chi* from the previous observation measurement. Again, move away 13 *zhang* 5 *chi* from the pole and observe the southern bank of the river's mouth; it is seen that [the bank] coincides with the pole in the south. What is the width of the river's mouth?

[Answer:] 1 *li* 200 *bu*.

[Method:] Multiply the [first] observation measurement [obtained] along the string by the final distance [of observation] from the pole and divide by the distance between poles. What is obtained is then subtracted from the initial distance [of observation] from the pole, giving the remainder as the *fa*. Then multiply the difference between the initial and final distances from the pole by the difference of the observation entries [obtained along] the string; the result is the *shi*. Dividing the *shi* by the *fa* yields the width by the mouth of the river.

[7] Now, for [the purpose of] looking into a deep abyss containing a clear pool of water with white stones at the bottom, hold a carpenter's square on the edge of the abyss. Assume that the height of the square is 3 *chi*. Sight downward [at an angle]; the water level is seen at 4 *chi* 5 *cun* along the lower base, while the white stone [at the bottom of the pool] is seen at 2 *chi* 4 *cun*. Erect another similar carpenter's square above [the first] such that it is 4 *chi* from the lower one and look obliquely downward again from the upper square; the water level is seen at 4 *chi* along the upper base, while the white stone is seen at 2 *chi* 2 *cun*. What is the depth of the water?

[Answer:] 1 *zhang* 2 *chi*.

[Method:] Place the observation measurements of the water level [obtained] along the upper and lower bases [on the counting board] and subtract one from the other. The remainder is then multiplied by the observation entry of the stone along the upper base. The result is taken as the upper rate. Obtain the difference of the observation measurements of the stone [taken along] the upper and lower bases and multiply the remainder by the observation measurement of the water level along the upper base. The result is taken as the lower rate. Subtracting one rate from the other and multiplying the remainder by the distance between the squares gives the *shi*. Multiply the two differences [obtained previously] together and take [this product] as the *fa*. Dividing the *shi* by the *fa* yields the depth of the water.

[8] Now, for [the purpose of] observing a river in the south from a mountain, erect a carpenter's square on the mountain. Assume that the height of the square is 1 *zhang* 2 *chi*. Look obliquely downward at the southern bank [of the river] from the tip of the square; the bank is seen at 1 *zhang* 3 *chi* 1 *cun* along the lower base. Sight again at the northern bank; it is seen at 1 *zhang* 8 *cun* within the previous observation entry. Climb up a hill [so that you move] 51 *bu* higher

and 22 *bu* to the north, set up the square [as before], and sight again at the southern bank along the tip of the square. The bank is seen at 1 *zhang* 2 *chi* along the base. What is the width of the river?

[Answer:] 2 *li* 102 *bu*.

[Method:] Multiply the [observation measurement] of the southern bank [obtained] along the lower base by the height of the carpenter's square and divide by the [observation measurement] along the upper base. Subtract the height of the square from the result obtained, thus giving the *fa*. Multiply the northward movement [placed on the counting board] by the height of the square and divide it by the [length of the] upper base. Subtract the results obtained from the upward movement and multiply the remainder by the difference of the observation measurements along the lower base, thus giving the *shi*. Dividing the *shi* by the *fa* yields the width of the river.

[9] Now, for [the purpose of] observing a city in the south from a mountain, set up a carpenter's square on the mountain. Assume that the upright arm [i.e., the height] of the square be 3 *chi* 5 *cun* and that the tip of the square is aligned with the southeastern corner of the city and with the northeastern corner. Sight at the northeastern corner from the tip of the upright arm; the corner is seen at a point 1 *zhang* 2 *chi* from the far end of the lower base. Then set the upright arm of the square down horizontally at the point of [the previous] observation measurement and sight at the northwestern corner from the tip of the base now standing; the corner is seen at a point 5 *chi* from the far end of the horizontal arm of the square. Sight on the southeastern corner [from the tip of the upright arm of the square in its original position]; the corner is seen at a point 1 *zhang* 8 *chi* from the far end of the lower base. Set up again another similar carpenter's square above, assuming that the two squares are 4 *zhang* apart. Sight on the southeastern corner from the tip of the upright arm; the corner is seen at a point 1 *zhang* 7 *chi* 5 *cun* from the far end of the upper base. What is the width and length of the city?

[Answer:] The length [of the city] in the north-south direction is

1 *li* 100 *bu*. The width in the east-west direction is 1 *li* 33 1/3 *bu*.

[Method:] Multiply the observation measurement of the south-eastern corner along the lower base by the height of the square and divide by its measurement along the upper base. Subtract the height of the square from what is obtained, thus giving the *fa*. Subtract the observation measurement of the northeastern corner along the lower base from that of the southeastern corner and multiply the remainder obtained by the distance between the squares, giving the *shi*. Dividing the *shi* by the *fa* yields the length [of the city] in the north-south direction. To find the width of the city, multiply the distance between the squares by the observation measurement along the upright arm of the square in horizontal position, giving the *shi*. Dividing the *shi* by the *fa* yields the width of the city in the east-west direction.

In future reference these problems are designated by contextual names in the following manner:

1. The sea island problem
2. The pine tree problem
3. Size of a distant walled city problem
4. Depth of a ravine problem
5. Height of a building as viewed from a hill problem
6. Width of river mouth problem
7. Depth of a clear pool problem
8. Width of a river problem
9. Size of a city observed from above problem

3

ANALYSIS AND DISCUSSION OF THE *HAIDAO*'S CONTENTS

The *Haidao* as an Extension of the *Jiuzhang*

The twenty-four problems of the "*Gougu*" chapter of the *Jiuzhang* serve as a primer of right triangle theory. The "Pythagorean proposition" is introduced, and its applications are reinforced by a variety of sometimes fanciful problem situations, involving vines winding about trees, logs encased in walls, broken bamboo shoots, and so on.[1] However, the last six problems clearly concern land surveying situations, such as finding distances to given objects or determining the size of walled villages. In particular, problem 23 involves finding the height of a far hill the distance

1. A complete translation and discussion of these problems can be found in Frank J. Swetz and T. I. Kao, *Was Pythagoras Chinese? An Examination of Right Triangle Theory in Ancient China* (University Park, Pa.: The Pennsylvania State University Press, 1977), See also, Lam Lay-Yong and Shen Kangsheng, "Right-Angled Triangles in Ancient China," *Archive for History of Exact Sciences* 30 (1984): 87–112.

to which is known, and problem 24 requires finding the depth of a well by using one sighting along a pole. Each of these surveying problems can be solved through the use of one sighting observation. A natural extension of this series of problems, both in mathematical complexity and in the realm of practical applications, is the determining of distances where two or more distinct observations are required. The most advanced mathematical technique required by the *Jiuzhang* problems involves using a simple proportion resulting from a similarity of right triangles. By contrast, all the addended *Haidao* problems require two or more observational sightings to resolve problem situations involving two sets of similar triangles where similarity holds within the sets but not between them. The similarity relationship gives rise to two sets of proportions that must be solved simultaneously, and this appears to be the focus of the *chong cha* solution technique.

Thus, both mathematically and theoretically with regard to surveying, the nine *Haidao* problems devised by Liu Hui were a natural and practical extension of the *Jiuzhang*.

A Surveying Thesis?

Although the *Haidao* contains only nine problems, their contents provide valuable insights into the status and conditions of early Chinese surveying. Clearly, the context of the problems affirms the theory that many early mathematics-based surveying activities were concerned with cartography and military affairs; problems 1, 2, 4, and 7 would lend themselves to topographical concerns, while the remaining problems—3, 5, 6, 8, and 9—have military significance. In particular, the "height of a building as viewed from a distance" problem could almost have been prefaced by a statement of Heron of Alexandria affirming the need for military surveying:

> How many times in the attack of a stronghold have we arrived at the foot of the ramparts and found that we made our ladders and other necessary implements for the assault too short, and have consequently been defeated simply for not knowing how to use the Dioptra [i.e., to use surveying] for measuring the heights of walls; such heights have to be measured out of the range of enemy missiles.[2]

---

2. Free translation of passages given in M. Vincent, "Le Traité de la Dioptre de Héron

The use of traditional instruments of surveying—sighting poles, the carpenter or set-square, and stretched ropes or strings—is obvious. Four problems require the use of sighting poles, and five employ applications of the set-square. The poles of 20 and 30 *chi*, approximately 7.0 and 10.5 meters respectively, were probably made of bamboo and could be easily transported and erected by two or three men. It should be remembered that surveying was not a solitary activity in the ancient world; it required a gang of laborers and assistants, besides the master surveyor. The standard Chinese set-square had legs of length 2 *chi*, but the set-square or gnomon described in Liu's "width of a river" problem is large, with legs of length 12 *chi* and 23+ *chi*. In an actual field situation, such squares were apparently to be constructed by attaching a horizontal member or beam to a standing vertical staff. In his translation of the "depth of a ravine" and "depth of a clear pool" problems, Ho chose to use this interpretation,[3] and later pictorial illustrations from Chinese surveying tracts indicate the use of such cross-staffs.

For accurate results in the sea island problem, the initial sightings have to be taken at the same ground level. Although the text mentions no leveling activities, level reference planes had to be established before observations were taken. This is apparently assumed, and there is no further elaboration on this subject.

Leveling was accomplished by the use of a water level, *chun*, which existed as far back as the Zhou dynasty (ca. eleventh century–221 B.C.), where, it is noted, "In construction, carpenters must make sure the ground is level."[4] Closer to the time of Liu Hui, Sui dynasty (581–618) documents, in addressing a specific construction project, note:

> Surveyors explain the methods of mathematics, make the land flat in the north-south side of the river, the land may be of a few hundred miles [in extent], make sure the north-south is level, use the plumb line for leveling.[5]

---

d'Alexandrie," *Notices et Extraits des Manuscrits de la Biblitothèque Impériale* 19 (1958), by R.C. Skyring-Walters, "Greek and Roman Engineering Instruments," *Transaction: The Newcomen Society for the Study of Engineering and Technology* 2 (1921): 45–60; Polybus (ca. 203–120 B.C.), a Greek historian who accompanied Scipio on his African campaigns, also noted the military importance of using astronomy (surveying) and geometry to find distances and heights of walls that could not be measured directly due to the enemies' presence.

3. Ho, *Dictionary of Scientific Biography*.

4. See Shen Kangsheng's discussion of surveying techniques in Wu Wenchun, ed., *Jiuzhang Suanshu yu Liu Hui*[cv] (Nine Chapters on the Mathematical Art and Liu Hui) (Beijing: Beijing Normal University, 1982), 9:181–90.

5. Ibid., p. 185.

The evidence offered by ancient maps shows that Chinese surveyors were able to obtain accurate orientations. Eight directions are specified among the *Haidao*'s problems: the four cardinal directions and northeast, northwest, southeast, and southwest. Proto-compasses were used in obtaining the directions.

Despite the informative nature of Liu's problems, there are some anomalies in the procedures described. In the sea island and pine tree problems, the observer awkwardly takes sightings from ground level rather than from eye level, and in the "depth of the pool" problem no allowances are made for the refractive index of water, thereby rendering the observations inaccurate. Liu Hui apparently was not a practicing surveyor. He was a scholar-mathematician intent on revising a standard reference and clarifying and strengthening the mathematics contained in it. The *Haidao Suanjing* eventually emerged as an independent surveying manual, but that was probably not Liu's intent.

The Origins and Concept of the *Chong Cha* Procedure

When Liu wrote his addendum to the "*Gougu*" chapter of the *Jiuzhang*, he called it "*chong cha,*" double difference," and noted that the solution method was not something new, but known and accepted at that time.[6] This solution method apparently evolved out of the Chinese concern for determining the solar distance, the distance from the earth to the sun. To accomplish this, shadow-reckoning with right triangle computational theory was employed. Early references to such attempts are made in *Zhouli*[cd] and in the Han astronomical manual *Huai Nan Zi*[ce] (The Book of [the Prince of] Huai Nan) (ca. 200 B.C.).[7] But the most detailed procedures were supplied in the *Zhoubi*, where it is noted that the shadow of an 8 *chi* gnomon would change 1 *cun* in length for each 1,000 *li* traversed in a north-south direction. Then a problem is investigated where if one travels 60,000 *li* from a point where the sun is directly overhead and erects an 8 *chi* gnomon, the gnomon's shadow would be 6 *chi*. By simple proportions, the

6. Qian Baocong, *Suanjing Shi Shu* (Ten Mathematical Manuals) (Shanghai: Zhoaghua Shuju, 1963), p. 92.

7. See discussion in Needham, *Science and Civilization in China*, pp. 224–25.

solar distance is found to be 80,000 *li*. Furthermore, the oblique distance from the foot of the shadow to the sun would be 100,000 *chi*, as found by using the *Gougu* (Pythagorean) theorem.[8] Zhang Heng[cf] (A.D. 78–139), in describing his cosmological theory in *Ling Xian*[cg] (Spiritual Constitution of the Universe), mentions the use of double right-angled triangles, the situation that gives rise to the *chong cha* method.[9]

Liu builds on this established interest and theory. In prefacing the *Haidao* problems, he illustrates the principle of "double differences" by generalizing the quantitative account of solar distance measurement given in the *Zhoubi*:

> Erect two gnomons at the city of Loyang. Let the height [of each gnomon] be eight *chi*. [Both the gnomons erected] in the north-south direction are on the same level. Measure the shadows [of the gnomons] at noon on the same day. The difference in length of the shadows is taken as the *fa*. Multiply the difference in distance of the gnomons by their height and take the result as the *shi*. Divide the *shi* by the *fa* and add to it the height of the gnomon. The result is the height of the sun from the earth. Take the shadow length of the southern gnomon and multiply it by the distance between the gnomons to give the *shi*. Divide the *shi* by the *fa*. The result is the distance of the southern gnomon from the subsolar point in the south at noon.[10]

Using modern techniques, the problem is easily analyzed. In Figure 5 the two gnomons are represented by $\overline{AS}$ and $\overline{CN}$, having equal lengths h. Let the distance SN between the gnomons be denoted by X, the length SB of the shadow of the southern gnomon be a_1, and the length ND of the shadow of the northern gnomon be a_2. At point C, construct $\overline{CE} \parallel \overline{AB}$ and $\overline{CR} \parallel \overline{QD}$. If the height of the sun is represented by Y, and the distance of the southern gnomon from the subsolar point Q is Z, then by using the pairs of similar triangles, PRA and CNE and PAC and CED, we obtain RP/NC = RA/NE = PA/CE = AC/ED or $(Y - h)/h = Z/a_1 = X/(a_2 - a_1)$, from which are found $Y = hX/(a_2 - a_1) + h$ and $Z = a_1X/(a_2 - a_1)$, where X is the difference in distance between the gnomons and $a_2 - a_1$ is the difference between the lengths of the gnomons' shadows. Hence, for

8. Li and Du, *Chinese Mathematics*, p. 30.
9. Needham, *Science and Civilization in China*, p. 104.
10. Qian, *Suanjing Shi Shu*, p. 92.

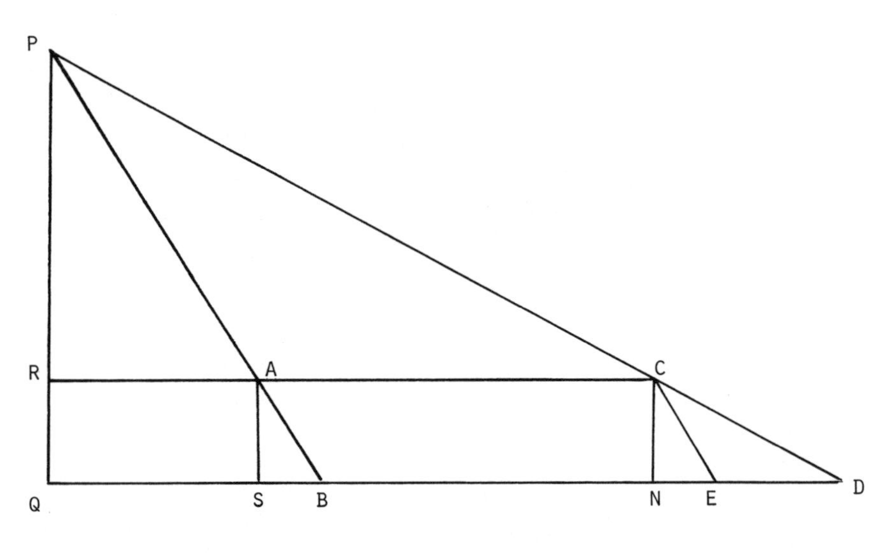

FIG. 5

finding the height and distance of an object from a point of observation, an expression of the "double difference" can be employed. Liu Hui obviously understood the principle of "double difference" and found that the method could be extended by using either double gnomons or set-squares to make three or four observations of inaccessible height, depth, or distance.[11]

Taking at face value the literal translation of the phrase "double difference," along with Liu's prefatory remarks and the solution formula for his first problem, one might infer that the solutions to the problems in the *chong cha* category require the securing and manipulation of two numerical differences, but this is not necessarily the case. The fact that not all the *Haidao*'s problems contain such numerical double differences in their solution schemes indicates that the term *chong cha* had a broader meaning. Thus, Mikami interpreted *chong cha* to mean repeated or double applications of proportions,[12] while van Hée, who doubted the literal meaning of "double difference," considered it to be an application of double pro-

---

11. Liu's extension of the "double difference" technique is the subject of a recent paper by Lih Ko Wei[cw] (Li Guowei), "From One Gnomon to Two Gnomons: A Methodological Study of the Method of Double Differences," presented at the Fifth International Conference on the History of Science in China, San Diego, California, August 1988.

12. Yoshio Mikami, *The Development of Mathematics in China and Japan* (1913; New York: Chelsea Publishing Co., 1974), p. 35.

portions.[13] Berezkina translated *chong cha* as "double-level differences," implying the use of physical measurements taken at two different levels or locations along the same line—a situation that would naturally result in the use of double proportions.[14] Even Chinese commentators questioned the scope of Liu's technical designations. In an examination of Liu's work, Yang Hui (A.D. 1275) noted that "men of the past changed the names of their methods from problem to problem."[15] If this is indeed the case, then at most the term *chong cha* can be assumed to designate a general class of problems concerning right triangles.

Mathematical Significance and Controversies

Within a modern mathematical perspective, it is easy to associate *chong cha* computational procedures with trigonometric thinking. Indeed, Alexander Wylie (1815–1887), the first Western commentator on the *Haidao*, described its contents to be "nine problems in practical trigonometry";[16] Joseph Needham, in a more recent view of the work, referred to *chong cha* as "a kind of empirical substitute for trigonometric functions."[17] The apparent trigonometric aspects of the problems deserve examination. Consider the modern verification of the solution for problem 3:

In Figure 6, GDOH is a square walled city. Sighting poles are placed at points F and C, and observations are taken from points A and B. Distances AB, BC, FC, and EC are determined. It is desired to obtain measures of the length of the city wall and the distance CD.

Solution:

(i) Construct a line from point E parallel to $\overline{GA}$ and intersecting $\overline{AO}$ at point J. In triangles FCA and ECJ, AC/JC = FC/EC implies JC = (AC)(EC)/FC; this is the "shadow difference." In triangles GBA and GDB, AB/JB = BG/BE and BG/BE = GD/EC, respectively. Combining these results, we see that AB/JB = GC/EC implies that

$$GD = \frac{(AB)(EC)}{JB}. \qquad (1)$$

13. Van Hée, "Le Classique de l'Île Maritime," p. 267.
14. Berezkina, *Matematika Drevnovo Kitaya*, p. 278.
15. Lam, *Critical Study*, p. 345.
16. Alexander Wylie, *Notes on Chinese Literature* (Shanghai, 1867).
17. Needham, *Science and Civilization in China*, p. 109.

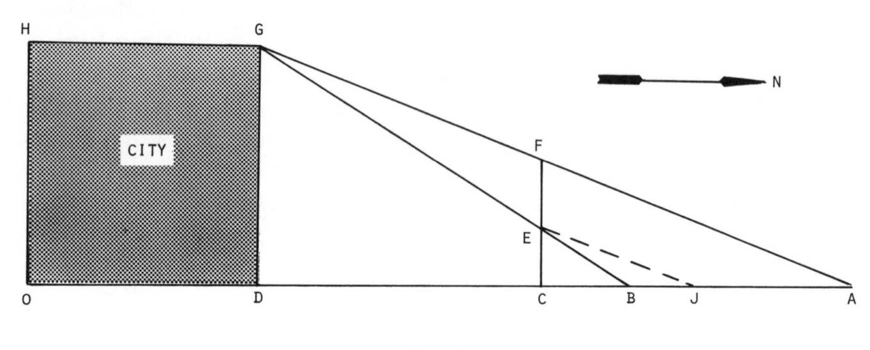

FIG. 6

Since AB = AC − BC and JB = JC − BC, we may substitute these results into (1) and obtain GD = (AC − BC)EC/(JC − BC).

(ii) In triangles GDB and GBA, CD/BC = EG/BE and EG/BE = AJ/JB, respectively. Combining these results, we see that CD/BC = AJ/JB and

$$\mathrm{CD} = \frac{(\mathrm{AJ})(\mathrm{BC})}{\mathrm{JB}}. \tag{2}$$

Since AJ = AC − JC and JB = JC − BC, substitution of these values into (2) yields CD = (AC − JC)BC/(JC − BC).

In this problem, Liu defines the "shadow difference," *jing cha*[de], to be JC − (AC)(EC)/FC. If in Figure 6, we let ∠ AFC = α, then tan α = AC/FC, and JC becomes the length of the projection of $\overline{\mathrm{EC}}$ along $\overline{\mathrm{AC}}$—that is, in a physical sense $\overline{\mathrm{JC}}$ may be described to be the shadow of $\overline{\mathrm{EC}}$ caused by light rays parallel to $\overline{\mathrm{GA}}$. This shadow difference ratio supplies a computational link between the two sets of similar triangles associated with this problem, and it facilitates a solution. While a modern student of mathematics might resolve the construction of $\overline{\mathrm{JE}}$ to establish a relationship between $\overline{\mathrm{EC}}$ and $\overline{\mathrm{FC}}$, Liu devised the ratio known as the "shadow difference" and implicitly employed the tangent function. Similar ratios appear in the solutions to problems 5, 6, 8, and 9. The use of "shadow functions"—the tangent and cotangent—were familiar to many ancient peoples who observed the heavens with the aid of a vertical staff, or gnomon.[18] Solutions of problems given in the *Gougu* section of the *Jiuzhang* also utilized ratios that today would be recognized as tangents of given angles; however, Liu's use of tangent ratios to solve intricate geometrical problems is far in

18. E.g., the Egyptians and the Babylonians.

advance of the procedures offered in the *Jiuzhang*, and his techniques, although limited, may be thought of as a prototrigonometry even though a concept of angle per se is not used.[19]

It is known that Liu, in his addendum to chapter 9 of the *Jiuzhang*, derived proofs for the solution formulas he gave, and those proofs were supplemented by diagrams.[20] This is extremely important, first because it distinguishes Liu Hui as the first known Chinese mathematician who sought proofs for his results and second because it raises the question as to the nature of his proofs. Mathematics, in and of itself, did not have a high scholarly priority in ancient China. It was considered a minor art whose application and results were necessary for the functioning of the empire and as such was not worthy of prolonged deliberation or extensive literary annotation. The terse mathematical explanations given in problem situations affirm this position. Thus, while satisfying the need to compile mathematical procedures that would ensure pragmatically correct results, Liu was also expressing himself as a mathematician by seeking out reasons for his results. The fact that the proofs were lost within a relatively short period of time also supports the impression that pure mathematics was not socially valued.

Because Liu's original proofs were lost, their actual form and theory are open to conjecture. Differing opinions as to the substance of these proofs have resulted in controversy. Even the great thirteenth-century mathematician Yang Hui was perplexed by "the setting of the methods and problems in the *Haidao*." When he wanted to include the *Haidao* problems in his *Xugu Zhaiqi Suanfa*, Yang had "to place a small diagram of the sea island problem before him so that he was able to understand a little of the method employed by his predecessors."[21] However, it is not clear whether Yang Hui was referring to an existing copy of Liu Hui's original diagram or to one he had constructed himself. From his deliberations on the problems, Yang did produce a theoretical proof of the *chong cha* method. More modern scholars working on reconstituted versions of the *Haidao* problems have tended to incorporate contemporary concepts, certainly unknown to Liu, into their solution verifications. For example, the explanations offered for "double difference" by Li Huang and Shen Qinpei in the early half of the nineteenth century depended on a geometry of similar triangles and contained many Euclidean constructions alien to ancient Chinese me-

---

19. For the ancient Chinese conception of angle, see Lih Ko Wei, "A Gestalt Interpretation of the Traditional Chinese Concept of Angle" (Paper presented at the Sixth International Conference on the History of Science in China, Cambridge, England, August 1990).

20. Ho, *Dictionary of Scientific Biography*.

21. Lam, *Critical Study*, p. 345.

thodology.[22] In his *Haidao Suanjing Wei Bi* (Notes on the Sea Island Mathematical Manual) of 1879, Li Liu employed the traditional "Celestial element" method, a Song-Yuan algebraic method derived subsequent to Liu's original work, to obtain solutions.

Since 1926, when Li Yan published his research on the origins and applications of the *chong cha* method, the text of *Haidao Suanjing* has attracted considerable attention among Chinese historians of mathematics.[23] Li Yan's exposition of Liu Hui's formulas were based on the properties of similar triangles, and for almost half a century this approach has remained the primary basis for Chinese research on the methods of the *Haidao*. Bai Shangshu[ch], in his recent study of the *Haidao* problems, surveyed historical explanations of the *chong cha* method as provided by Chinese mathematicians from the Song-Yuan period (A.D. 960–1368) onward.[24] On the basis of this survey, Bai is convinced that Liu Hui used a theory of proportions involving similar right triangles to derive proofs for the *Haidao* formulas. Wu Wenjun[ci], however, opposes this theory and offers a different geometric insight to interpret Liu Hui's work.[25] Convinced that ancient Chinese mathematicians did not employ explicit concepts of angles or parallel lines in their theories, Wu bases his understanding of Liu's methods on an analysis of Zhao Shuang's[cj] "diagram of solar height" given in the *Zhoubi Suanjing*. He speculates that Liu, in studying this diagram, had perceived that the complements of the angles formed by the diagonals of a given rectangle are pairwise congruent, citing Liu Hui's commentary to chapters 5 and 9 of the *Jiuzhang Suanshu* as evidence for this hypothesis. Further, Wu asserts that the "small diagram of the sea island," which Yang Hui referred to in his *Xugu Zhaiqi Suanfa*, may well have been a copy of Liu Hui's original diagram.[26] Thus, Wu believes

22. The ancient Chinese did not have a theory of parallel lines that allowed for possible construction of auxiliary lines in a geometric problem situation, nor did they employ a general theory of proportionality based on corresponding parts of similar triangles. Their use of proportionality was restricted to similar right triangles. See C. N. Liu[cx] (Liu Caonan), "Haidao Suanjing Yuanliu Kao"[cy] (An Investigation into the Origins of the Sea Island Mathematical Manual), *Yi Shi Bao Wen Shi Fu Kan*[cz], no. 21 (1942).

23. Li Yan, "Chong Cha Shu Yuanliu Ji Qi, Xin Zhu."

24. S. S. Bai, "Liu Hui Suanjing Zao Shude Tantao"[da] (Investigation and Discussion of Methods Used in Liu Hui's Sea Island Mathematical Manual), *Kojishi Wenji*[db] 8 (1982): 79–87.

25. Wu Wenchun, "Wo Guo Gudai Cewang Zhi Xue Chongcha Lilun Pingjji Jian Ping Shuxueshi Yanjiu Zhong Mo Xie Fangfa Wenti"[dc] (Commentary on the Value of the Theory of *Chong Cha* in the Study of Ancient Chinese Measurement and Observation and the Commentary on Certain Questions in the Research of the History of Mathematics), *Kojishi Wenji* 8 (1982): 10–30.

26. Ibid., p. 13.

that his reconstructed geometric proofs for the *Haidao Suanjing* formulas are similar to those Yang Hui suggested in the thirteenth century.

The Methodology Behind Liu's Proofs

In reading the solution instructions to the *Haidao*'s problems, one is struck by their algebraic nature, yet the situations acted on are primarily geometric. As with many early peoples, Chinese mathematical methodology was intuitively based and usually bound to a visual perception. Arithmetic and geometric concepts could be frequently interchanged—for example, the product of two numbers a and b could be conceived of as the area of a rectangle with sides of lengths a and b, and if the numbers were equal their product could be visualized as the area of a square. In deriving proofs, the Chinese utilized this geometric perception combined with dissection techniques. Their methods depended on the use of two assumptions:

1. If a plane figure is dissected into several regions, the sum of the areas of the separate regions is equal to the area of the plane figure.
2. Area remains invariant under a set of rigid transformations in the plane.

These assumptions became the basis of an "out-in complementary principle" whereby Chinese mathematicians rearranged geometric shapes to substantiate arithmetic properties.[27] It appears that this principle was used in arriving at the theory of proportions evident in such works as the *Zhoubi* and the *Jiuzhang*. A brief illustration of the "out-in" methodology in this case is as follows:

---

27. Wu Wenchun, "The Out-In Complementary Principle," in *Ancient China's Technology and Science*, pp. 66–89. This type of geometric-algebraic thinking was not unique to the Chinese. Evidence of such solution procedures exists in Babylonian writings and in Greek mathematical works. The Chinese "out-in complementary principle" applies to situations involving rectangles and depends on the fact that the complements of rectangles about a diagonal of a given rectangle are equal in area. This concept was formalized by Euclid as Proposition 43 in Book I of his *Elements* and used extensively by early Greek mathematicians. The Chinese geometric-algebraic techniques do appear to differ from those of the Greeks in that the Chinese apparently had no theory of parallel lines that allowed for the use of auxiliary lines in geometric problem-solving, and their use of the properties of geometric similarity was restricted to working with right triangles. In general, it appears that Chinese mathematical solution techniques were more rigid than those of the Greeks.

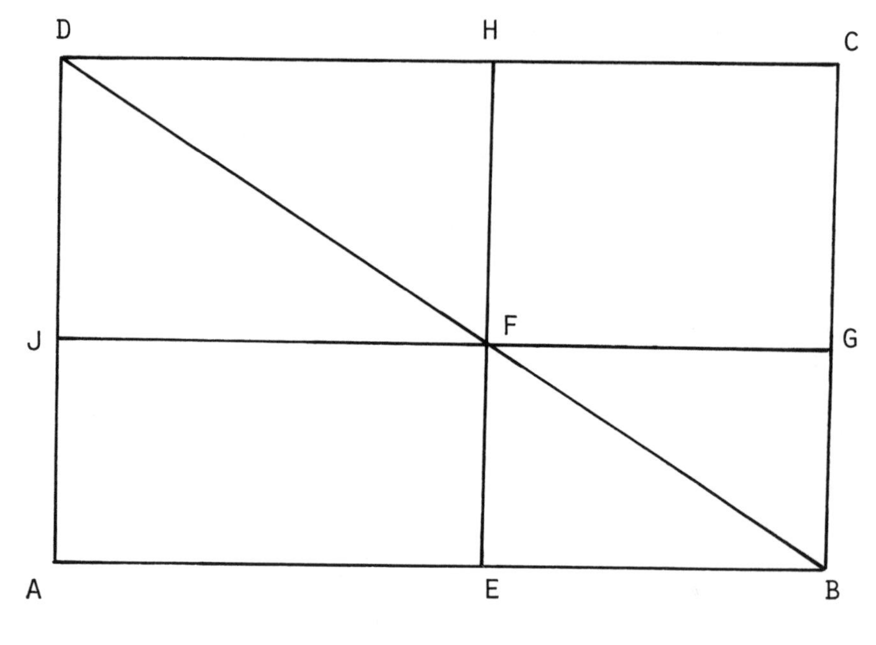

FIG. 7

Consider □ AC as dissected in Figure 7.[28]

Since a diagonal draw to a rectangle bisects the rectangle, a(△DAB)[29] = a(△BCD), and similarly a(△FEB) = a(△BGF) and a(△DJF) = a(△FHD); therefore a(□ AF) = a(□ FC). It then follows that:

$$AE \times EF = FG \times GC, \; AB \times BG = EB \times BC, \ldots$$

$$JF : EB = DJ : FE, \; AB : EB = DA : FE, \ldots$$

A Euclidean interpretation would derive these proportions on the basis of corresponding parts of similar triangles—that is, △DJF ~ △FEB, △DAB ~ △FEB..., but through the use of the out-in principle the concept of similarity is avoided. Derivations for Liu's *chong cha* procedure were apparently also derived through a use of "out-in complementary principle"

28. Rectangles will be designated by a diagonal, the diagonal that extends from the lower left vertex to the upper right vertex—i.e., AC indicates rectangle ABCD. These conventions will be used throughout the discussion.

29. To be read as "area of triangle DAB."

thinking. The evidence for such a conclusion, while speculative, is fairly substantial.

The first known commentary on the *Zhoubi* was produced in about the third century by Zhao Shuang. In his work, Zhao attempted to justify the solution expressions obtained for the solar height problem. He drew a "diagram of solar height" supplemented with colored regions of blue and yellow on which he applied the out-in principle. The surviving diagram attributed to Zhao (see Figure 8a) has suffered under the interpretations of successive nonmathematical copyists. Qian Baocong has reconstructed the diagram according to the instructions in Zhao Shuang's commentary—with one small liberty: the addition of an auxiliary line from the tip of the rear pole (see Figure 8b).[30] His explanation of the diagram is as follows:

> [The rectangles called] Yellow A and Yellow B are actually equal in area. Yellow A is the product of the height of the pole and the distance between the poles. The difference between the lengths of the shadows forms the width of Yellow B, so when [the area] is divided by this, the result obtained is the length of Yellow B, its upper end being on the same level as the sun. According to the diagram the height of a pole has to be added to this. [I] say [it is] 80,000 *li*, from the pole added to the top. [The rectangles called] Blue C and Blue F are also equal in area. When Yellow A and Blue C are joined together and Yellow B and Blue F are also joined together, their combined areas are equal.[31]

A closer analysis and reinterpretation of these statements using modern notation reveals both the power of the "out-in complementary principle" and the facility with which ancient Chinese mathematicians must have employed it. Consider the "solar height" diagram in Figure 5 altered to a rectangular configuration, as shown in Figure 9.

In $\square QG$, by the out-in principle, $a(\square QA) = a(\square AG)$; in $\square QF$, $a(\square QC) = a(\square CF)$. Rectangle AG is translated horizontally to the right so that A maps onto C; thus, $a(\square AG) = a(\square CM)$. Under Zhao's scheme, $\square QA$ = Blue C, and $\square CM$ = Blue F, thus Blue C = Blue F. Now, $a(\square QC) = a(\square QA) + a(\square SC)$, $a(\square CF) = a(\square CM) + a(\square OF)$ as $\square SC$ = Yellow A and $\square OF$ = Yellow B, then Yellow A = Yellow B. Employing the previously used substitutions for relevant measurements—that is, $PQ = Y$,

30. Qian, *Suanjing Shi Shu*, p. 24.

31. As translated in Wu Wenchun, "Commentary on the Value of the Theory of *Chong Cha*" (1982), p. 11.

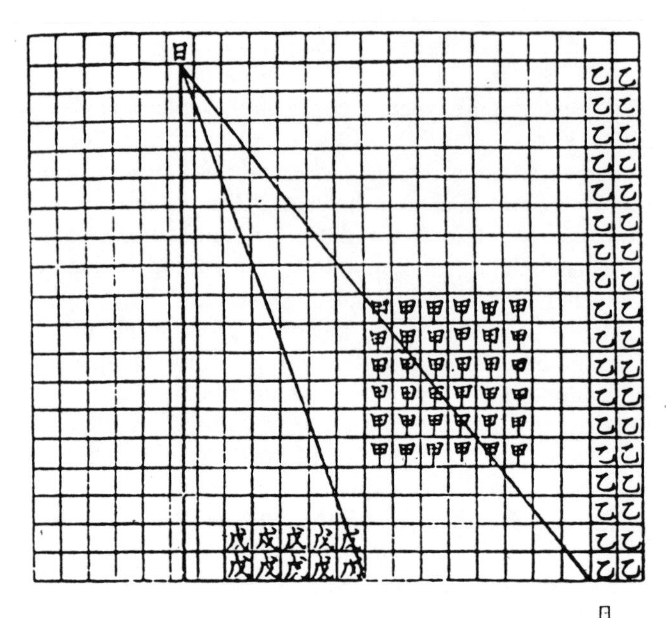

(a)

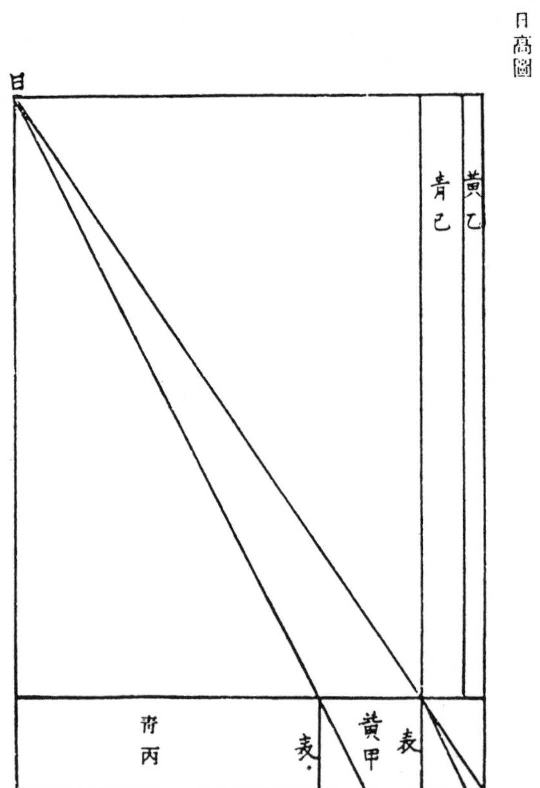

(b)

FIG. 8

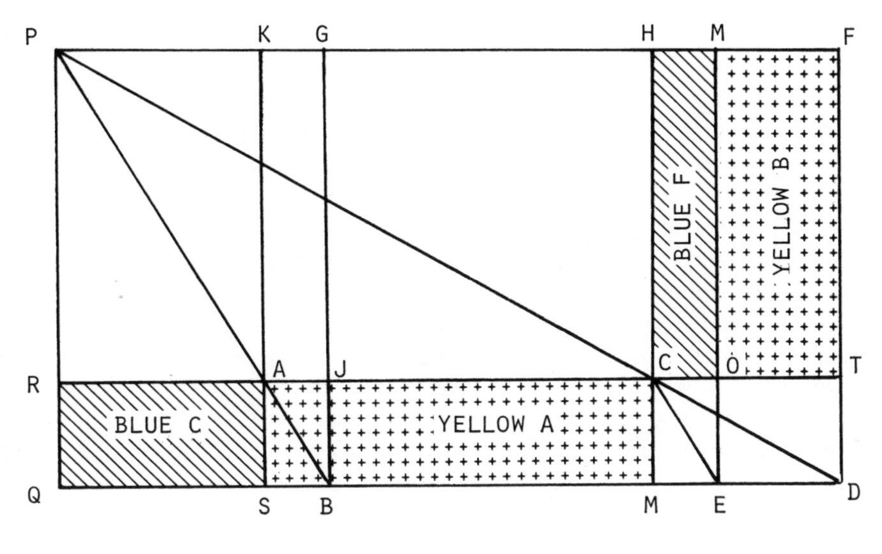

FIG. 9

SN $=X$, AS $=h$, SB $=a_1$ and ND $=a_2$, the algebraic interpretation of the results supplies:

$$hX=(PR)(a_2-a_1) \Rightarrow PR=\frac{hX}{a_2-a_1} \text{ and } Y=\frac{hX}{a_2-a_1}+h.$$

Since Liu Hui and Zhao Shuang were contemporaries, this technique was certainly known to both. Liu wrote about such geometric experimentation and manipulation in his commentary to the *Jiuzhang*, where he advised: "Use diagrams on small papers, cut diagonally through the intersections, rearrange them together, combine them to form particular shapes. . . ."[32] In his thirteenth-century commentary on the sea island problems, Yang Hui used the same geometric-based method of derivation. Both tradition and the stagnation of mathematical thought had served to preserve this approach to mathematical proof over the intervening years.

32. Li and Du, *Chinese Mathematics*, p. 71.

Verification of the *Sea Island Manual* Formulas

Since the time of Yang Hui, various attempts have been made to justify mathematically the validity of the *Sea Island Manual* formulas. Some most recent attempts include those of E. I. Berezkina (1974), Bai Shangshu (1982), and Wu Wenjun (1982). Bai contends that Liu's proofs were devised by use of simple proportions, whereas Wu theorizes that the out-in complementary principle served as a basis for the ancient proofs. Lam Lay-Yong[ck] and Shen Kangsheng[cl] have categorized the *Haidao* problems into groups whose formulas can be proven by the use of four traditional methodologies.[33] While the exact methods Liu used to obtain proofs will remain unknown, it is apparent that there were several traditional methodologies through which he could have obtained proofs. His reliance on the use of diagrams and commentary on the necessity of constructing paper models and moving pieces would seem to support the theory that his proofs depended on a use of the out-in principle. The following verification will take this approach while also utilizing the theory of simple proportion evidenced in the *Jiuzhang* and a method of analogy—that is, reinterpreting new situations in terms of those already known and solved.

1. *The sea island problem*

In Figure 10, AS, CN, SN, SB, and ND are known; PQ represents the height of the island; and QS is the distance from the island to the first sighting pole $\overline{AS}$. Observations are taken from the points B and D.

Referring to Figure 10 and applying the out-in principle:

$$a(\square QC) = a(\square CF)$$

$$a(\square QA) = a(\square AJ) = a(\square CI)$$

$$a(\square SC) = a(\square QC) - a(\square QA) = a(\square CF) - a(\square CI).$$

$$\text{Thus, } a(\square SC) = a(\square MF)$$

$$\text{or } (AS)(SN) = (PR)(ME) = PR(ND - SB)$$

$$\text{and } PR = \frac{(AS)(SN)}{(ND - SB)},$$

33. Lam Lay Yong and Shen Kangsheng, "Mathematical Problems on Surveying in Ancient China," *Archive for History of Exact Science* 32 (1986): 1–20.

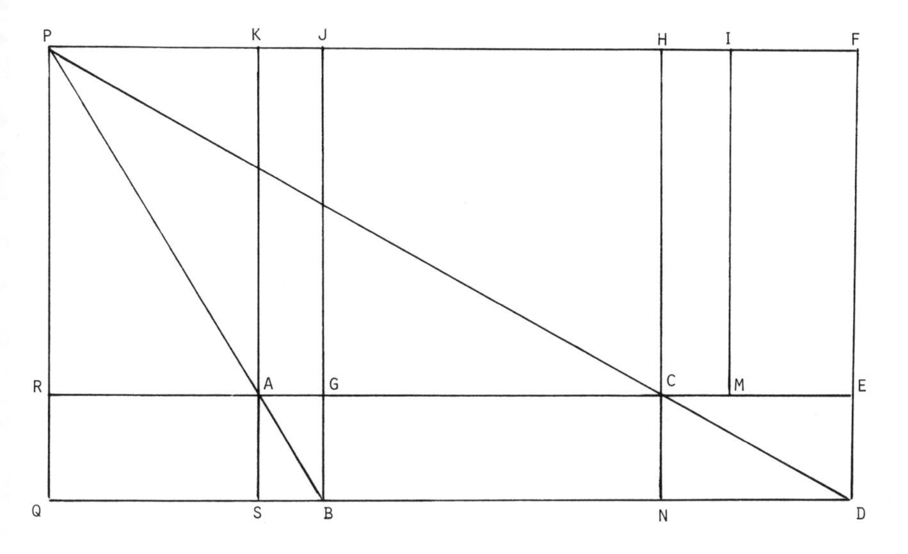

FIG. 10

from which it follows that

$$\boxed{PQ = \frac{(AS)(SN)}{(ND - SB)} + AS.}$$

Also, since a (□ QA) = a (□ CI),

$$(QS)(AS) = (SB)(PR)$$

$$\text{or} \quad (QS)(AS) = (SB)\frac{(AS)(SN)}{(ND - SB)},$$

$$\text{and} \quad \boxed{QS = \frac{(SB)(SN)}{(ND - SB)}.}$$

2. *The pine tree problem*

In Figure 11, let AB be the height of the tree and DK the distance of the hill from the nearest sighting pole; $\overline{EF}$ and $\overline{CD}$ are sighting poles of known height. Observations are taken from points H and G, and distances HF, GD, FD, and CL are determined.

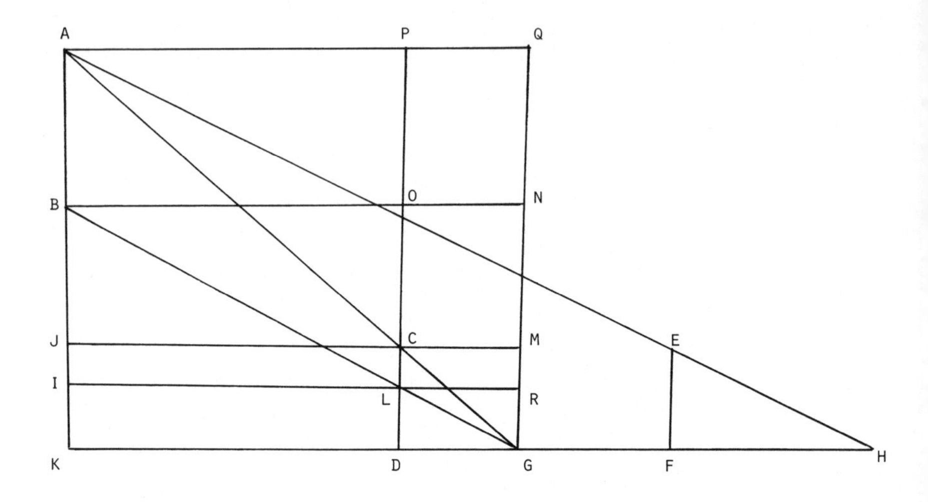

FIG. 11

A direct application of the distance formula developed in the sea island problem results in the fact that

$$\boxed{DK = \frac{(GD)(FD)}{(HF - GD)}.}$$

When the out-in principle is applied to the regions given in Figure 11, it can be shown that:

$$a(\square CQ) = a(\square KC)$$
$$a(\square LN) = a(\square KL)$$
$$a(\square IC) = a(\square KC) - a(\square KL).$$

Thus, $a(\square IC) = a(\square CQ) - a(\square LN)$

and $(CL)(DK) = (CM)(QM) - (CM)(OL)$

$= (CM)(QM - OL).$

Since is it seen that $(QM - OL) = (QN - CL)$, and $QN = AB$, $CM = GD$, this last expression may be rewritten as:

$$(CL)(DK) = GD(AB - CL)$$

or

$$AB = \frac{(CL)(DK)}{GD} + CL.$$

Substituting the expression for DK obtained above into this formula, it is found that

$$\boxed{AB = \frac{(CL)(FD)}{(HF - GD)} + CL.}$$

3. *Size of a distant walled city problem*

In Figure 12, GDOH is a square walled city. Sighting poles are placed at points F and C, and observations are taken from points A and B. Distances AB, BC, FC, and EC are determined. It is desired to obtain measures of the length of the wall and the distance CD.

By the out-in principle it is found that:

$$a(\square CM) = a(\square KI)$$

or $(EC)(CA) = (FC)(KA)$,

from which can be found

$$KA = \frac{(EC)(CA)}{FC}.$$

Now, by analogy, if one considers $\overline{EC}$ and $\overline{JK}$ to be sighting poles and $\overline{CB}$ and $\overline{KA}$ their respective "shadows," the problem situation becomes the same as that of the sea island problem, and the appropriate formulas apply:

$$CD = \frac{(BC)(CK)}{(KA - BC)}.$$

Since $CK = CA - KA$,

$$\boxed{CD = \frac{BC\,(CA - KA)}{(KA - BC)}.}$$

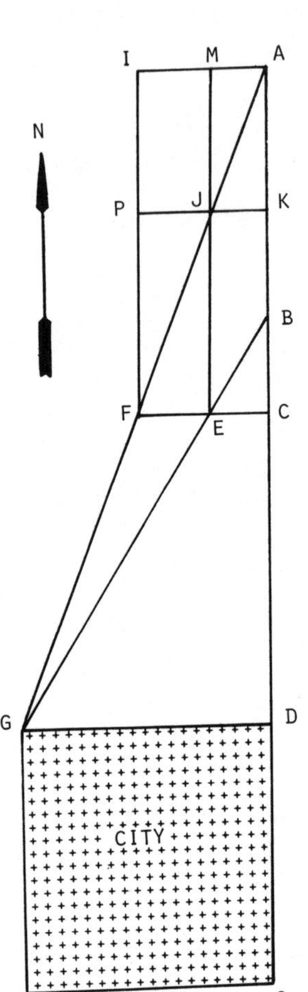

FIG. 12

Similarly,

$$GD = \frac{(EC)(CK)}{(KA - BC)} + EC$$

$$\text{or } GD = \frac{(EC)(CA - KA)}{(KA - BC)} + EC.$$

Thus

$$\boxed{GD = \frac{(EC)(AC - BC)}{(KA - BC)}.}$$

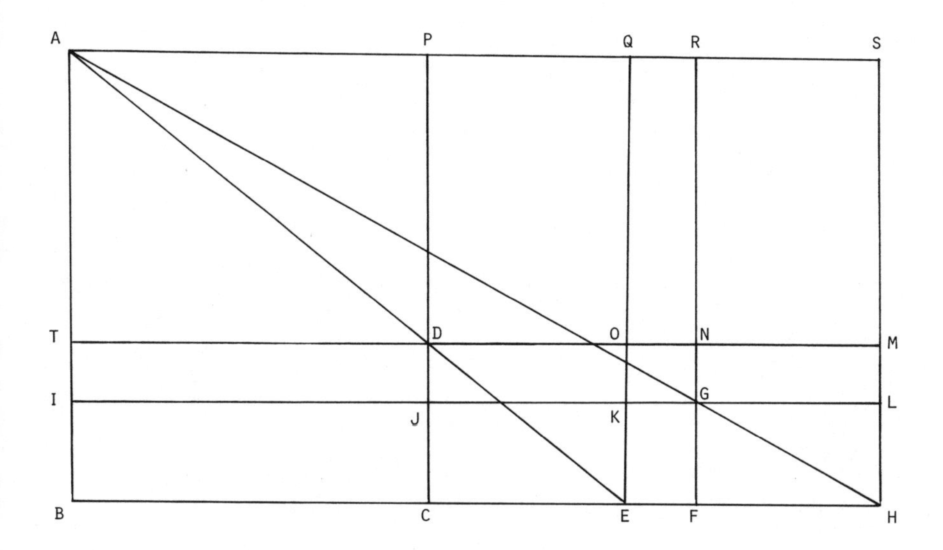

FIG. 13

4. *Depth of a ravine problem*

In Figure 13, let CB be the unknown depth of the ravine. Carpenter squares are set up on points C and F, and sightings are taken from points H and E along the extended legs of the respective carpenter squares. The distances HF, GF, FC, EC, and DC are known.

When the out-in principle is applied on the regions shown in Figure 13, it is found that:

$$a(\square GS) = a(\square BG)$$

$$a(\square NS) = a(\square GS) - a(\square GM).$$

$$\text{Thus, } a(\square NS) = a(\square BG) - a(\square GM) \qquad (3)$$

$$a(\square ID) = a(\square BD) - a(\square BJ)$$

$$\text{but } a(\square BD) = a(\square DQ) = a(\square NS).$$

$$\text{Therefore, } a(\square ID) = a(\square NS) - a(\square BJ).$$

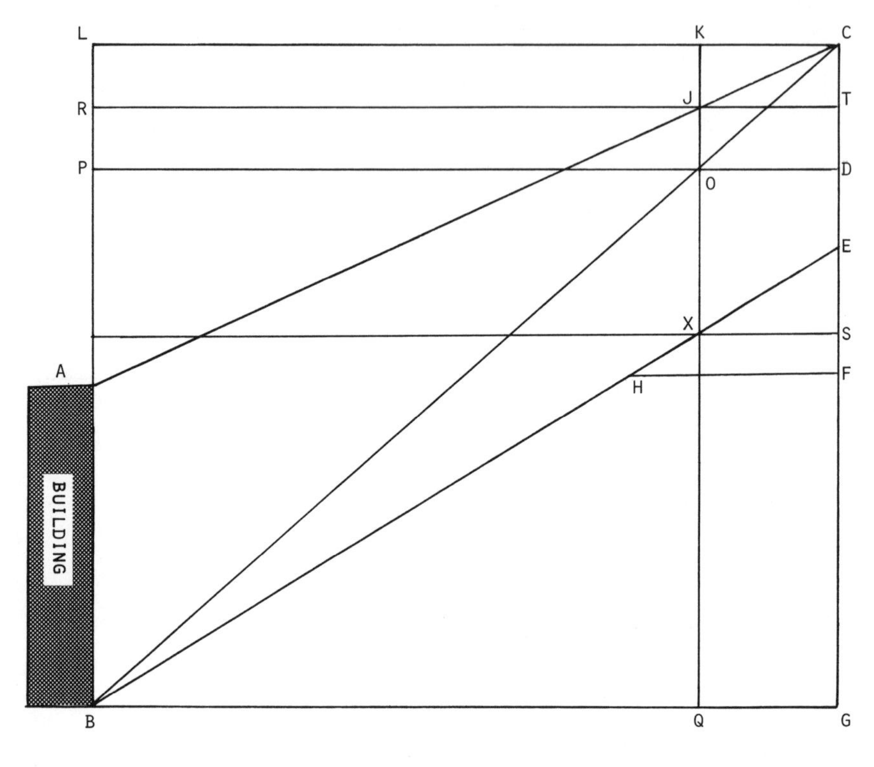

FIG. 14

Substituting the expression obtained for a (□ NS) in equation (3) into this last result, we obtain:

$$a(\square\, ID) = a(\square\, BG) - a(\square\, GM) - a(\square\, BJ).$$

Since $a(\square\, BG) - a(\square\, BJ) = a(\square\, CG)$

$$a(\square\, ID) = a(\square\, CG) - a(\square\, GM).$$

Thus, $(BC)(DC - GF) = (GF)(CF) - (FH)(DC - GF)$

and $$\boxed{BC = \frac{(GF)(CF)}{(DC - GF)} - FH.}$$

5. *Height of a building as viewed from a hill problem*

In Figure 14, let AB be the unknown height of the building. Sightings are taken from points C and E, and the distances CD, EF, JO, OD, HF, and DF are known.
Applying the out-in principle, the following relationships appear:

$$a(\square PK) = a(\square RK) + a(\square PJ) \Rightarrow a(\square PJ) = a(\square PK) - a(\square RK)$$

$$\text{but } a(\square RK) = a(\square XT) \quad \text{and} \quad a(\square PK) = a(\square QD)$$

$$a(\square PJ) = a(\square QD) - a(\square XT).$$

$$\text{Thus, } (JO)(BQ) = (OD)(DG) - (OD)(TS)$$

$$= OD(DG - TS).$$

$$\text{Since } (DG - TS) = (SG - TD) = (AB - JO),$$

$$\text{then } (JO)(BQ) = OD(AB - JO)$$

$$\text{and } AB = \frac{(JO)(BG)}{OD}. \tag{4}$$

In this expression, BG remains unknown but

$$BG = BQ + QG = BQ + OD$$

$$\text{since } a(\square PK) = a(\square QD),$$

$$(CD)(BQ) = (OD)(DG) \Rightarrow BQ = \frac{(OD)(DG)}{CD}.$$

$$\text{Thus, } BG = \frac{(OD)(DG)}{CD} + OD = \frac{OD(DG + CD)}{CD} = \frac{(OD)(CG)}{CD}.$$

Substituting this expression for BG in equation (4) above, we obtain

$$AB = \frac{(JO)(OD)(CG)}{(OD)(CD)}.$$

But $CG = CF + FG$. Therefore,

$$AB + \frac{(JO)}{(CD)}(CF + FG).$$

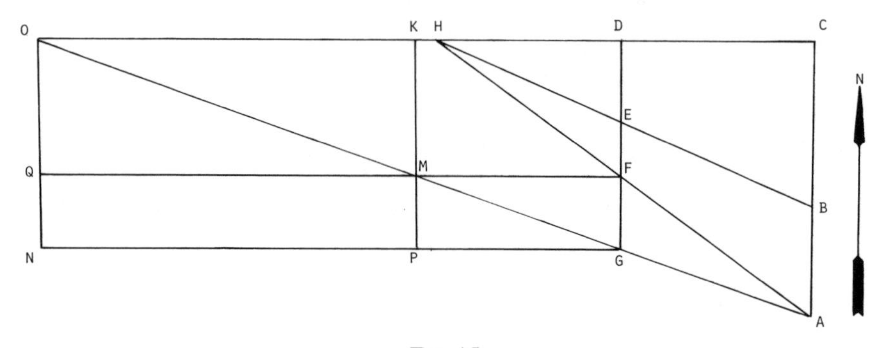

FIG. 15

FG may be found by using the ravine formula. Thus:

$$FG = \frac{(DF)(OD)}{HF - OD} - EF .$$

$$\text{Now } AB = \frac{(JO)}{(CD)} \left[CF + \frac{(DF)(OD)}{HF - OD} - EF \right]$$

$$\text{or } AB = \frac{(JO)}{(CD)} \left[\frac{(CF - EF)(HF) + (EF - CF + DF)OD}{HF - OD} \right].$$

Since $(CF - EF) = CE$ and $EF = CD$,

$$\boxed{AB = \frac{(CE)(HF)(JO)}{(HF - OD)\,EF}}$$

6. *Width of a river mouth problem*

In Figure 15, let AB represent the unknown width of the river mouth. Sighting poles are set inland at points D and G, and the distances HD, DF, EF, OD, and DG are known. Consider this problem a variation of the pine tree problem, where the width of the river AB is the height of the pine tree, and $\overline{DF}$ is a sighting pole. Let $\overline{KM}$ also be a sighting pole where $KM = DF$.

Then:

$$a(\square\ MD) = (\square\ NM)$$

or $(FD)(DK) = (GF)(KO)$.

Since $DK = DO - KO$,

$$(FD)(DO - KO) = (GF)(KO) \Rightarrow KO = \frac{(FD)(DO)}{GD}. \qquad (5)$$

Using the pine tree formula, we obtain:

$$AB = \frac{(FE)(DO - DH)}{KO - DH} + FE = \frac{(FE)(DO - DH)}{KO - DH}.$$

But the value for KO can be given by equation (5). Therefore,

$$\boxed{AB = \frac{(FE)(DO - DH)}{\frac{(FD)(DO)}{GD} - DH}.}$$

7. *Depth of a clear pool problem*

In Figure 16, let AB represent the unknown depth of the water. B marks the position of the white stone, F indicates the edge of the abyss, and AG is the surface of the water. Observations are taken from points E and C determining OD, JD, KF, and LF; CD, DF, and FG are also known.

This problem can be solved through the use of two applications of the depth of a ravine formula. Apply the depth formula to find FH and then FG. The depth of the clear pool $AB = FH - FG$.

$$FH = \frac{(DF)(JD)}{(LF - JD)} - CD$$

$$FG = \frac{(DF)(OD)}{(KF - OD)} - CD$$

$$AB = \frac{(DF)(JD)}{(LF - JD)} - CD - \frac{(DF)(OD)}{(KF - OD)} - CD.$$

Thus, $AB = \frac{(DF)[JD(KF - OD) - OD(LF - JD)]}{(LF - JD)(KF - OD)}$.

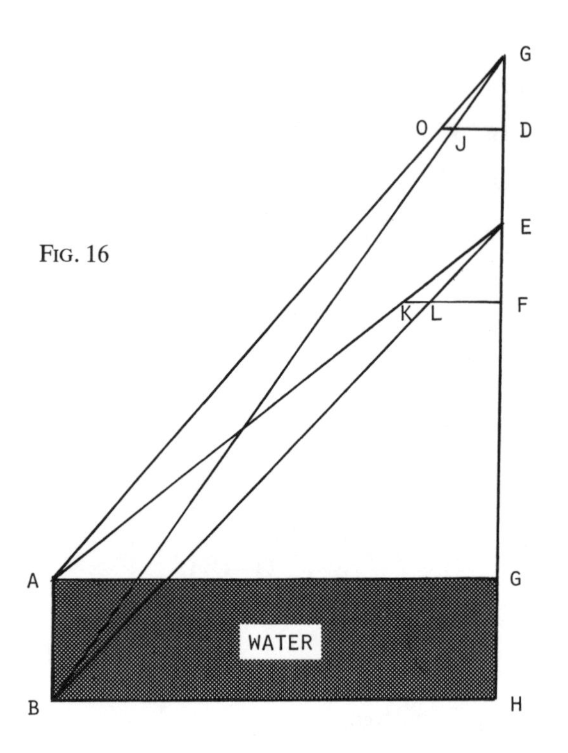

FIG. 16

8. *Width of a river problem*

In Figure 17, let AB represent the unknown width of the river, and LK and PG 'the heights of the sighting squares, LK = PG. Observations are taken from points P and L determining the distances JK, KH, GH, EG, and EF. This problem can be considered a variation of the pine tree problem and the pine tree formula can be adapted to obtain a solution. In Figure 17, NQ can be considered an additional pole, and then by the pine tree formula:

$$\mathrm{AB} = \frac{(\mathrm{EF})(\mathrm{HO})}{\mathrm{QR} - \mathrm{GP}} + \mathrm{EF} \tag{6}$$

$$= \frac{\mathrm{EF}(\mathrm{HO} + \mathrm{OR} - \mathrm{GP})}{\mathrm{QR} - \mathrm{GP}}.$$

Consider the numerator, where GP = KL, HO + QR = HS, and HS = HL − LS.
Therefore, (HO + QR − GP) = (HL − LS − KL) = (HK − LS).

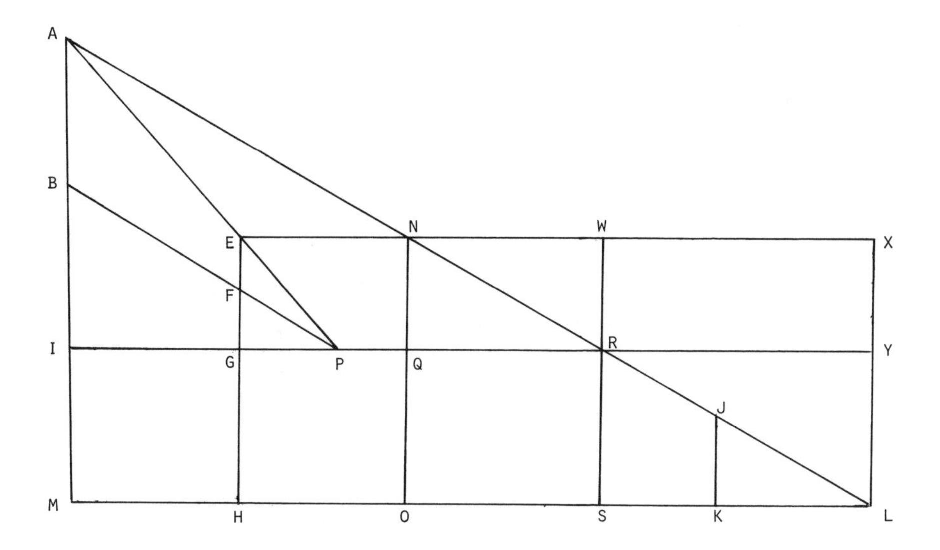

FIG. 17

Now $\triangle JKL \sim \triangle RSL \Rightarrow \dfrac{LS}{RS} = \dfrac{LK}{JK}$, and since $RS = GH$,

$$LS = \frac{(LK)(GH)}{JK} \tag{7}$$

The numerator for expression (6) now becomes

$$EF\left[HK - \frac{(LK)(GH)}{JK}\right]. \tag{8}$$

The denominator of expression (6) is $QR - GP = OS - LK$, and because $a(\square RX) = a(\square OR)$, by the out-in principle; it follows that

$$(LS)(EG) = (GH)(OS) \Rightarrow OS = \frac{(LS)(EG)}{GH}.$$

Furthermore, by using expression (7),

$$OS = \frac{(LK)(GH)(EG)}{(JK)(GH)} = \frac{(LK)(EG)}{JK},$$

and, finally, $QR - GP = \dfrac{(LK)(EG)}{JK} - LK.$

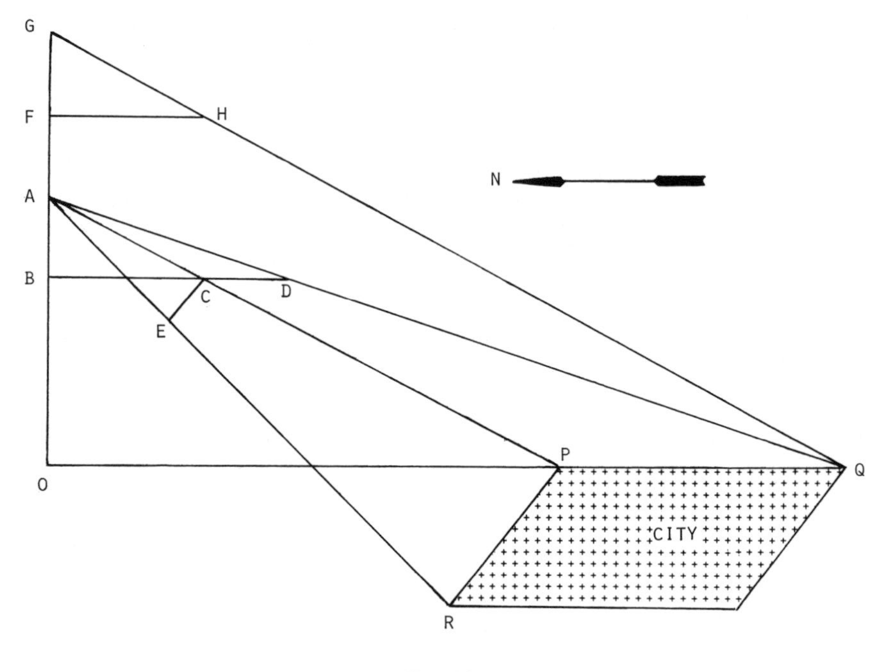

FIG. 18

Substitute the new numerator and denominator into the expression given by formula (6), and

$$AB = \frac{EF\left[HK - \frac{(GH)(LK)}{JK}\right]}{\frac{(LK)(EG)}{JK} - LK}.$$

9. *Size of a city observed from above problem*

In Figure 18, PQ is the unknown length of the city and PR is the unknown width. Observations are taken from points A and G on a distant mountain from which distances AB, BC, CE, BD, FH, and FB are either known or determined.

The length of the city, PQ, can be found by using the pine tree formula, where PQ is assumed to be the height of the tree:

$$PQ = \frac{(CD)(BO)}{AB} + CD = \frac{CD\,(BO + AB)}{AB} = \frac{(CD)(AO)}{AB}.$$

Now AO = BO + AB and AB = GF.

Using the "depth of a ravine" formula,

$$BO = \frac{(FB)(FH)}{BD - FH} - GF.$$

$$\text{Thus, } AO = \frac{(FB)(FH)}{BD - FH} \tag{9}$$

$$\text{and } PQ = \frac{(CD)(FB)(FH)}{AB(BD - FH)},$$

which is equivalent to the given solution formula:

$$\boxed{PQ = \frac{FB(BD - BC)}{\frac{(BD)(AB)}{FH} - AB}.}$$

$$\triangle APR \sim \triangle ACE \Rightarrow \frac{AP}{AC} = \frac{PR}{EC} \Rightarrow PR = \frac{(AP)(EC)}{AC}$$

$$\triangle ABC \sim \triangle AOP \Rightarrow \frac{AP}{AC} = \frac{AO}{AB} \text{ and } AP = \frac{(AO)(AC)}{AB}.$$

$$\text{Thus, } PR = \frac{(AO)}{(AB)}(EC),$$

where the value for AO is obtained from (9) above.

$$\text{Thus, } \boxed{PR = \frac{(FB)(FH)(EC)}{(AB)(BD - FH)}.}$$

4

CONCLUSIONS

A Mathematical Testimony

The contents of the *Haidao Suanjing* supply valuable exposure to one form of mathematical application—surveying—in ancient China and to the skill and genius of its originator, Liu Hui. It also provides insights into the evolution of Chinese mathematical thought and supplies a link of understanding between the practices of shadow-reckoning and scientific observation based on proven and accepted mathematical principles.

Liu was most judicious in initiating his study of *chong cha* by merely referencing his techniques to the "solar height problem" but then demonstrating these techniques within the realm of terrestrial surveying. Astronomical observations and results were of primary importance to the Imperial Court, but Liu realized from the beginning the spatial limitations of *chong cha*: the results were valid only for relatively short distances measured along a plane. The traditional belief that the shadow of an 8 *chi* gnomon would change 1 *cun* in length for each 1,000 *li* transversed in a north-south direction was based on the standing cosmological assumption of a flat earth and was false.[1] Liu must have realized this and avoided controversy and confrontation by proceeding as he did. Liu's problems were all practically conceived. Their results concerning sizes and distances—for example, the height of a building, 19 meters (problem 5), and the length of a city wall, 1.3 kilometer (problem 3)—are accurate as supported by historical evidence. In its time the *chong cha* technique was good; it worked and became a basis for further mathematical investigations. The astronomer-mathematician Li Chunfeng modified *chong cha* observations to be applicable on an inclined plane, thereby broadening the scope of the technique.[2] Li also refuted the 1 *cun* per 1,000 *li* shadow ratio theory and challenged the flat earth hypothesis. In A.D. 724 the State Astronomical Bureau of the Tang dynasty initiated the first meridian survey in the ancient world. Under the supervision of the scholar Yixing[df], thirteen observation stations were established near the meridian 114°E and between latitudes 29°N to 52°N.[3] Using standard 8 *chi* gnomons and shadow-reckoning techniques, observations were taken over several years. The expedition determined the angular attitude of the north celestial pole above the horizon and recorded the lengths of shadows at noon for the summer and winter solstices and equinoxes. The differences in the shadow lengths found were much shorter than the supposed ratio of 1 *cun* per 1,000 *li* and disproved the traditional belief.[4] *Chong cha* techniques, although tried, were found unsatisfactory for astronomical purposes. In this work with shadows, Yixing devised and used formal tables of values for the tangent functions of given angles, providing yet another first for Chinese mathematics.[5]

By advancing from survey observations with one gnomon entailing the use of a simple ratio, to observations with two or more gnomons and

---

1. See discussion on page 41 above. Traditional Chinese cosmology held to the theory of a flat earth. Although some early schools of cosmology did propose a spherical earth, it was not until the second half of the seventeenth century that this theory was formalized and accepted.

2. Lih, "From One Gnomon to Two Gnomons."

3. For more details of the event, see Bo Shuren,[dd] "Astrometry and Astrometric Instruments," in *Ancient China's Technology and Science*, pp. 15–32; the work of Yixing is discussed in Ang Tian Se, "The Astronomical System of Yixing (A.D. 683–727)," *Bulletin of Chinese Studies, University of Hong Kong* 1 (1987): 1–12.

4. Yixing determined that one degree in polar elevation corresponded to 351 *li* 80 *bu*. For a discussion of this work, see Ang Tian Se, "A Biography of Yixing (683–727)," *Kertas-Kertas Pengajian Tionghoa* 3 (December 1989): 31–76.

5. China can be credited with many firsts in the history of mathematics, including use of decimal fractions, negative numbers, and accurate root extraction procedures. For the tangent table hypothesis, see C. Cullen, "An Eighth-Century Chinese Table of Tangents," *Chinese Science* 5 (1982): 1–33.

multiratio computations, Liu was taking a radical mathematical step. While the scope of shadow-reckoning activities was thereby vastly broadened, the mathematical complexity of the resulting solution formulas also became more intense. The solution procedures for the *Haidao* problems were more intricate than those of any of their predecessors in the *Jiuzhang* collection. Because *Haidao* procedures were not algorithmic, and their rationales were not intuitively obvious, Liu devised geometric-based verifications for his solutions. He proved his results and established a precedent for later Chinese mathematicians to prove and/or improve the *Haidao* techniques. The first known commentator on Liu's extended *Jiuzhang*, Zu Congzhi (429–500), also included proofs in his commentary (A.D. 450), but these soon became lost. Mathematical interest in the *Haidao* problems was particularly evident during the Song dynasty. Qin Juishao (ca. 1202–1261) improved on the problems by lifting the observation levels used from ground level to 5 *chi*, the normal height of an adult Chinese—even though he really didn't understand the functioning of the problems and made several serious mistakes in their interpretation. Qin's contemporary Yang Hui (fl. 1261–1275) was both perplexed and intrigued by the Sea Island Manual problems: "The subjects of the original treatise cover a very wide scope and are too difficult to apply or prove."[6] He criticized Li Chunfeng for not supplying justifications for the solutions in his previous commentaries (i.e., ca. 656), and he set out to devise his own proofs. Yang believed that "Liu Hui formulated his nine *Haidao* problems by employing the art of using other essentials to transform the technique of *chong cha* into a method of dealing with subtracted areas."[7] In brief, the formulas were derived using the "out-in complementary principle." Yang then employed this principle in developing proofs. Apparently the resulting proofs were highly admired and preserved, appearing three hundred years later in Cheng Dawei's[cm] *Suan Fa Tong Zong*[cn] (Systematic Treatise on Arithmetic) (1592). This evidence of a tradition of proofs should dispel the myth that among early societies the Greeks alone employed deductive demonstrations to ensure mathematical validity. The Chinese developed and used proofs from very early times.

Under the assumption that the ordering of the extant *Haidao* problem collection approximates Liu's original sequence, both a logical and purposeful mathematical and pedagogical systemization is evident in the problem listing. The first, sea island problem introduces and develops the *chong cha* strategy. Then the following problems consider a series of varied

6. Lam, *Critical Study*, p. 180.
7. Ibid., p. 14.

situations in which *chong cha* can be applied. The physical geometry of these situations encourages perceptual acuity and requires an extended use of analogy to the first problem. Liu made it quite clear that the required problem-solving should be accomplished through analogy:

> For measuring height, two rods should be set up, and for measuring depth a set of gnomons is needed. If an additional surveying point is necessary, we must observe three times; and if the surveying points are not on the same level, we must make four observations. In comprehending by analogy, problems are always solvable, though they may be very troublesome and obscure. I hope the learned readers will study the work in detail.[8]

Within the corpus of Chinese mathematical literature, the *Haidao Suanjing* is certainly a classic.

Chinese Surveying Accomplishments: A Comparative Retrospection

Problems involving right triangles and their uses in shadow-reckoning appeared in ancient China as early as 1000 B.C.[9] The *Haidao Suanjing*, with its nine surveying problems and right triangle computations, is part of this tradition. For more than a millennium the *Haidao*'s influence was apparent in the work of Eastern mathematicians.[10] Although it is a product of third-century China, a comparison of its contents and methodology with those of similar works or activities outside China helps establish its significance within the history of mathematical achievement.

Liu initiated his discussion of measuring inaccessible distances by considering the solar height problem from the *Zhoubi Suanjing*. While the

8. Qian, *Suanjing Shi Shu*, p. 92.

9. See Ang Tian Se, "Chinese Interest in Right-Angled Triangles," *Historia Mathematica* 5 (1978): 253–66; B. S. Gillon, "Introduction, Translation, and Discussion of Chao Chun-Ch'ing's 'Notes to the Diagrams of Short Legs and Long Legs of Squares and Circles,'" *Historia Mathematica* 4 (1976): 253–93.

10. In the seventh century, the *Haidao Suanjing* was adopted in Japan and Korea as part of the formal mathematics curriculum. *Haidao* problems appeared in the work of later Indian mathematicians. The transmission of the sea island problem from the East to the West is discussed in Kurt Vogel, "Ein Vermessungsproblem Reist von China nach Paris," *Historia Mathematica* 10 (1983): 360–67.

dating of the *Zhoubi* remains controversial, it is probably safe to conclude that the inclusion of such problems indicates that Chinese astronomers were concerned with finding the solar distance prior to the second century B.C. It is interesting to note that, about this same time, the Greeks made similar attempts to measure the heavens. Aristarchus of Samos (ca. 310–230 B.C.) determined the height of the sun employing right triangle ratios. Like his Chinese contemporaries, he made false assumptions and obtained inaccurate results.

While there is no doubt that the Greeks used mathematics effectively in surveying their structures and cities, few extant records of their practices exist. The primary source of information on land surveying practice in the ancient Western world is supplied by Heron of Alexandria's *Dioptra* (ca. A.D. 62). Heron compiled manuals of information on varied mathematical and scientific subjects. The formats and contents of his surviving manuals indicate that he was more concerned with practical applications of the subject rather than with theoretical explanations.[11] The initial chapters of the *Dioptra* describe the appearance and construction of a surveying instrument of the same name, a prototype theodolite. This instrument provided an accurate means of taking sightings and determining angles of elevation.[12] In the remainder of his book, Heron goes on to describe particular applications of the dioptra, including:

1. Determining the breadth of a river with access to only one side
2. Obtaining the distance between two inaccessible points
3. Determining the height to an inaccessible point and the difference in height between two inaccessible points
4. Obtaining the depth of a ditch through observation[13]

While the problems in these applications are similar to some of those considered by the Chinese in the *Jiuzhang Suanshu* and the *Haidao Suanjing*, their solution methods are primarily instrumental in nature and depend on a correct use of the instrument rather than on mathematical formulation and inventiveness. The mathematics of Greek surveying as presented by the *Dioptra* center around a use of simple proportions based

11. See A. G. Drachmann, "Hero of Alexandria," in *Dictionary of Scientific Biography*, 6: 310–14.

12. Skyring-Walters, "Greek and Roman Engineering Instruments."

13. A German translation of Heron's *Dioptra* is available: Hermann Schöne, *Heron Alexandrinus*, 3 vols. (Leipzig: B. G. Teubner, 1903). Perhaps the best English-language discussion of Heron and his work can be found in Thomas Heath's *History of Greek Mathematics*, vol. 2 (1921; New York: Dover Publications, 1981).

on the similarity of triangles. This evidence makes it appear that although the Greeks were more proficient in the technical art of surveying, the Chinese exceeded them in the application of mathematics to surveying situations.

Greek and Etruscan mathematical surveying methods were adopted and modified by the Romans, who used them in military campaigns and in the settlement of conquered territories. The Romans compiled surveying manuals that were broad in scope and included many aspects of surveying practice, physical, legal, and theoretical. These manuals included an introduction to applied geometry.[14] With the fall of the Roman Empire and the breakdown of its institutions, surveying as a recognized and respected activity almost ceased to exist. The societal and political stimuli that encouraged mathematical surveying simply did not exist during the Middle Ages. These stimuli, which were apparent during the Greek and Roman eras of Western dominance and continued in the Chinese empire, were:

1. A need for mapping associated with the existence and maintenance of an imperial or political state
2. Confirmation of land boundaries for purposes of private ownership and taxation assessment
3. Construction of public works, such as roads, aqueducts, and canals
4. Siege warfare

With the decline of feudalism and the rise of mercantile capitalism, these stimuli returned to the European scene, and practical geometry became valued once again.[15] The French scholar and cleric Gerbert (940–1003), who later became Pope Sylvester II, acquired a background in geometry by studying Arabic sources in Spain (967–970) and while he was abbot of the monastery of Bobbio in northern Italy, where he had available a *Corpus Agrimensorium* in the monastery library. Gerbert's *Gerberti Isagoge Geometriae* became the first medieval European work that included practical geometric applications among which were geometric situations and solution processes similar to those discussed in the *Haidao*. This work was followed by Savasorda's (ca. 1070–1130) *Liber Embodorum*, a text that

---

14. See O.A.W. Dilke, "Illustrations from Roman Surveyors' Manuals," *Imago Mundi* 21 (1967): 9–29; F. Blume et al., *Die Schriften der römischen Feldmesser*, 2 vols. (Hildesheim: Olms, 1967).

15. Mercantile capitalism was a major impetus to the resurgence of mathematical interest during the early Renaissance. A more complete discussion of this phenomenon is given in Frank J. Swetz, *Capitalism and Arithmetic: The "New Math" of the Fifteenth Century* (La Salle, Ill.: Open Court, 1989).

was compiled from Spanish sources and that dealt principally with geometric applications in finding lengths, heights, and areas; there were also practical geometries, such as that of Hugonis.[16] However, it was the appearance of Leonardo of Pisa's *Practica Geometriae* (1220) that gave surveying a firm mathematical basis.[17] Leonardo's work was strongly influenced both by Gerbert's *Geometriae* and by Plato of Tivoli's 1145 translation of the *Liber Embadorum*. Instruments such as the quadrant and the astrolabe became familiar to European surveyors, and the concept of angle as a mathematical entity emerged.[18] But despite these advances, it was not until the widespread use of artillery as a siege weapon that distance measure to inaccessible objects became an important subject in surveying manuals and problems began to take on a complexity similar to that found in the *Haidao*. See Figure 19.

When Xu Guangqi collaborated with the Jesuit Matteo Ricci in the seventeenth century to publish *Ce Liang Fa Yi* (Essentials of Surveying Trigonometry) (1607–1608), the Chinese were introduced to the methods and mathematics of European surveying, but the fifteen problems in that work did not equal or exceed the mathematical sophistication of the *Haidao Suanjing*. It is interesting to note that the tenth problem was the same as the sea island problem, but solved using the Euclidean mathematics of similar triangles.[19]

From the review and examination of early Western sources on applied geometry and land surveying, it appears that Chinese efforts in adapting mathematical principles to surveying situations surpassed those achieved in the West until the time of the European Renaissance.[20] Quite simply, in the endeavors of mathematical surveying, China's accomplishments exceeded those realized in the West by about one thousand years.

---

16. See R. Baron, "Hugonis de Sancto Victore *Practica Geometriae*," *Osiris* 12 (1956): 176–224.

17. Leonardo's *Practica* consists of eight chapters, two of which, chapters 3 and 7, are specifically devoted to the needs of surveyors. In chapter 3 Leonardo describes instrumental methods and computational techniques involving the use of a Ptolemaic chord table; chapter 7 discusses the use of a quadrant in angle measurement and the determination of inaccessible distances, and develops rules of surveying based on a similarity of triangles.

18. Surveying became a trigonometric activity. See M. C. Zeller, *The Development of Trigonometry from Regiomontanus to Pitiscus* (Joliet, Ill.: Sisters of St. Francis of Mary Immaculate, 1946). For an excellent discussion of early European surveying instruments and their uses, see E. R. Kiely, *Surveying Instruments: Their History and Classroom Use* (New York: Teachers College Press, 1947).

19. Ricci's proof is reproduced in Wu Wenchun, "Haidao Suanjing Gu Zheng dao Yuan" (Investigations into the Original Proofs of the Sea Island Mathematical Manual), in *Jiuzhang Shanshu* (1982) pp. 162–81.

20. A good survey of European surveying at this time is given in S. K. Victor, *Practical Geometry in the High Middle Ages* (Philadelphia: American Philosophical Society, 1979).

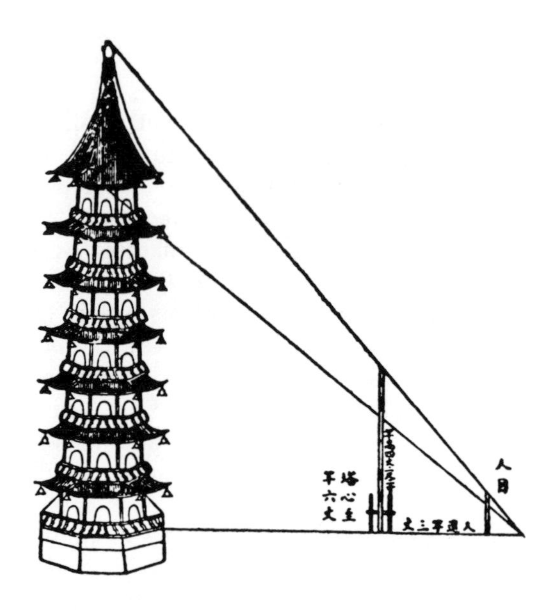

(a)

LIBER PRIMVS. 20

alter acutus F B A. Quare per sextam primi elem: duo latera eis opposita F A, A B erunt æqualia. Quod demonstrandum erat.

PROPOSITIO XI.

Eandem distantiam diametralem signi in plano positi à signo quopiã in altum sito ædificij perpendiculariter ad illud planum erecti, ita tamen, ut & ad signum plani, & ad basim ædificij accedi possit, per Quadratum Geometricum indagare.

Poteris vero per quadratum Geometricum distantiam diametralem A B in hunc modum venari. Posito Quadrato ad perpendiculum obseruabis signum B per dioptram, & notabis punctum ab ea sectum, quod quidem erit vel in latere D E vmbræ rectæ, vel in linea A E vmbræ mediæ, vel deniq; in latere E F vmbræ versæ.

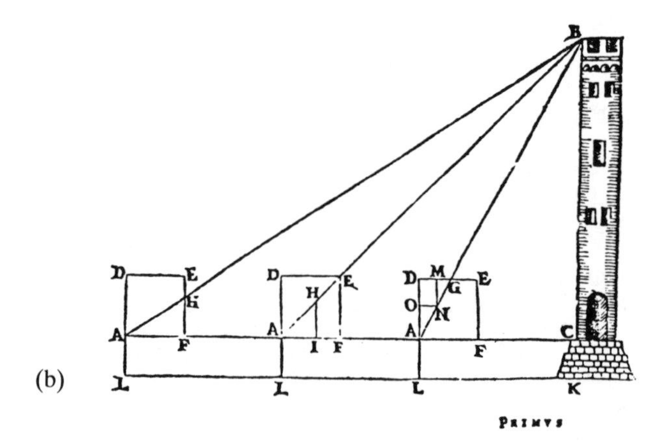

(b)

FIG. 19. Tower problems. (a) Reprinted from Qin Juishao, *Shu Shu Jiu Zhang* (+1247). (b) Reprinted from G. A. Magini, *De Planis Triangulis, Liber Unicus* (1592).

GLOSSARY

a 周禮
b 戰國策
c 管子
d 孫子兵法
e 孫臏兵法
f 裴秀
g 伏羲
h 女媧
i 禹
j 表
k 矩
l 縣
m 準
n 步車
o 規
p 望筒
q 照盤
r 周髀算經
s 周公
t 商高
u 海島算經
v 周
w 髀
x 疇人
y 九章算術
z 勾股
aa 劉徽
ab 重差
ac 祖沖之
ad 張邱建
ae 張邱建算經
af 算經十書
ag 李淳風
ah 秦九韶
ai 數書九章
aj 楊輝
ak 續古摘奇算法
al 田畝比類乘除捷法
am 朱世傑
an 田元玉鑑
ao 四元術
ap 永樂大典
aq 利瑪竇
ar 徐光啟
as 測量法義
at 崇禎曆書
au 大測
av 測量全義
aw 測量異同
ax 康熙
ay 測量高遠儀器用法
az 梅穀成

ba 數理精蘊
bb 戴震
bc 四庫全書
bd 孔繼涵
be 屈曾發
bf 武英殿
bg 曹操
bh 孫權
bi 劉備
bj 曹丕
bk 鄧艾
bl 李潢
bm 海島算經細草圖説
bn 沉欽裴
bo 重差圖説
bp 李鏐
bq 海島算經緯筆
br 天元
bs 李儼
bt 錢寶琮
bu 何丙郁
bv 洪天賜
bw 法
bx 實
by 里
bz 尺
ca 丈
cb 步
cc 寸
cd 周禮
ce 淮南子
cf 張衡
cg 張憲
ch 白尚恕
ci 吳文俊
cj 趙爽
ck 藍麗容
cl 沈康身
cm 程工位
cn 算法統宗
co 曹婉如
cp 杜石然
cq 正教奉褒
cr 郭書春
cs 重差術反其源流
ct 學藝
cu 梅榮照
cv 九章算術與劉徽
cw 李國偉
cx 劉操南
cy 海島算經源流考
cz 益世報文史副刊

da 劉徽《海島算經》造術的探討
db 科技史文集
dc 我國古代測望之學重差理論評介兼數學數研究中某些方法問題
dd 薄樹人
de 景差
df 一行

REFERENCES

Amma, T.A.S. *Geometry in Ancient and Medieval India*. Delhi: Motilal Banarsidass, 1979.

Ang Tian Se. "Chinese Computation with the Counting-Rods." *Kertas-Kertas Pengajian Tionghoa* 1 (1977): 97–109.

———. "Chinese Interest in Right-Angled Triangles." *Historia Mathematica* 5 (1978): 253–66.

———. "The Astronomical System of Yixing (A.D. 683–727)." *Bulletin of Chinese Studies, University of Hong Kong* 1 (1987): 1–12.

———. "A Biography of Yixing (683–727)." *Kertas-Kertas Pengajian Tionghoa* 3 (1989): 31–76.

Ang Tian Se and Frank J. Swetz. "A Chinese Mathematical Classic of the Third Century: *The Sea Island Mathematical Manual* of Liu Hui." *Historia Mathematica* 13 (1986): 99–117.

Bai, S. S. "Liu Hui Suanjing Zao Shude Tantao." *Kojishi Wenji* 8 (1982): 79–87.

Baron, R. "Hugonis de Sancto Victore *Practica Geometriae*." *Osiris* 12 (1956): 176–224.

Berezkina, E. I. "Drevnekitaishu Traktat Matematika v Devyati Knigakh." *Istoriko-Matematicheskie Issledovaniya* 10 (1957): 423–584.

———. "Dva Teksta Lyu Khueya Po Geometrii." *Istoriko-Matematicheskie Issledovaniya* 19 (1974): 231–48.

———. *Matematika Drevnovo Kitaya*. Moscow: Nauka, 1980.

Blume, F., et al. *Die Schriften der römischen Feldmesser*. 2 vols. Hildesheim: Olms, 1967.

Bo Shuren. "Astrometry and Astrometric Instruments." In Institute of the History of Natural Sciences, Chinese Academy of Science, *Ancient China's Technology and Science*, pp. 15–32. Beijing: Foreign Languages Press, 1983.

Bond, J. D. "The Development of Trigonometric Methods Down to the Close of the XVth Century." *Isis* 4 (1921): 295–323.

Cao Wanru. "Maps 2,000 Years Ago and Ancient Cartographical Rules." In Institute of the History of Natural Sciences, Chinese Academy of Science, *Ancient China's Technology and Science*, pp. 250–57. Beijing: Foreign

Languages Press, 1983.
Cullen, C. "An Eighth-Century Chinese Table of Tangents." *Chinese Science* 5 (1982): 1–33.
Dilke, O.A.W. "Illustrations from Roman Surveyors' Manuals." *Imago Mundi* 21 (1967): 9–29.
———. The Roman Land Surveyors. New York: Barnes & Noble, 1971.
Drachman, A. G. "Hero of Alexandria." In *Dictionary of Scientific Biography*, 6: 310–14. New York: Scribner's, 1970–80.
Gillon, B. S. "Introduction, Translation, and Discussion of Chao Chun-Ch'ing's 'Notes to the Diagrams of Short Legs and Long Legs of Squares and Circles.'" *Historia Mathematica* 4 (1976): 253–93.
Goodrich, Carrington L. *A Short History of the Chinese People*. New York: Harper & Row, 1959.
Guo Shuchun. "The Numerical Solution of Higher Equations and the *Tianyuan* Method." In Institute of the History of Natural Sciences, Chinese Academy of Science, *Ancient China's Technology and Science*, pp. 111–23. Beijing: Foreign Languages Press, 1983.
Heath, Thomas. *A History of Greek Mathematics*. 2 vols. 1921. New York: Dover Publications, 1981.
Hedquist, Bruce E. "On the History of Land Surveying in China." *Surveying and Mapping* 35 (September 1975): 251–54.
Ho Peng Yoke. "Liu Hui." In *Dictionary of Scientific Biography*, 8:418–25. New York: Scribner's, 1973.
———. *Li, Qi, and Shu: An Introduction to Science and Civilization in China*. Hong Kong: Hong Kong University Press, 1985.
Hoe, Jock. *Les Systèmes d'Équations Polynômes dans le Siyuan Jujian (1303)*. Mémoires de l'Institut des Hautes Études Chinoises, vol. 6. Paris: Collège de France, 1977.
Institute of the History of Natural Sciences, Chinese Academy of Science. *Ancient China's Technology and Science*. Beijing: Foreign Languages Press, 1983.
Karpinski, Louis C. "Roman Surveying." *School Science and Mathematics* 26 (1926): 853–55.
Kiang, T. "An Old Chinese Way of Finding the Volume of a Sphere." *Mathematical Gazette* 56 (1972): 88–91.
Kiely, E. R. *Surveying Instruments: Their History and Classroom Use*. New York: Teachers College Press, 1947.
Lam Lay-Yong. *A Critical Study of the Yang Hui Suan Fa*. Singapore: University Press, 1977.
Lam Lay-Yong and Shen Kangsheng. "Right-Angled Triangles in Ancient China." *Archive for History of Exact Sciences* 30 (1984): 87–112.
———. "The Chinese Concept of Cavalieri's Principle and Its Application," *Historia Mathematica* 12 (1985): 219–28.
———. "Mathematical Problems on Surveying in Ancient China." *Archive for History of Exact Sciences* 32 (1986): 1–20.
Lane, E. W. "Ingenuity of the Ancient Chinese." *Civil Engineering* 1 (1930): 17–22.
Latourette, Kenneth Scott. *The Chinese, Their History and Culture*. New York: Macmillan, 1964.
Li Yan. "Chong Cha Shu Yuanliu Ji Qi Xin Zhu." *Xue Yi* 7 (1926): 1–15.
Li Yan and Du Shiran. *Chinese Mathematics: A Concise History*. Translated by John N. Crossley and Anthony Lun. Oxford: Clarendon Press, 1987.
Libbrecht, Ulrich. *Chinese Mathematics in the Thirteenth Century: The Shu-Shu*

Chiu-Chang of Ch'in Chiu-Shao. Cambridge, Mass.: MIT Press, 1973.

Lih Ko-wei (Li Guowei). "From One Gnomon to Two Gnomons: A Methodological Study of the Method of Double Differences." Paper presented at the Fifth International Conference on the History of Science in China, San Diego, California, August 1988.

———. "A Gestalt Interpretation of the Traditional Chinese Concept of Angle." Paper presented at the Sixth International Conference on the History of Science in China, Cambridge, England, August 1990.

Liu, C. N. (Liu Caonan). "Haidao Suanjing Yuanliu Kao." *Yi Shi Bao Wen Shi Fu Kan*, no. 21, 1942.

Mei Rongzhao. "The Decimal Place: Value Numeration and the Rod and Bead Arithmetics." In Institute of the History of Natural Sciences, Chinese Academy of Science, *Ancient China's Technology and Science*, pp. 57–65. Beijing: Foreign Languages Press, 1983.

Mikami, Yoshio. *The Development of Mathematics in China and Japan*. 1913. New York: Chelsea Publishing Co., 1974.

Needham, Joseph. *Science and Civilization in China*, vol. 3. Cambridge: Cambridge University Press, 1959.

Neuberger, A. *The Technical Arts and Sciences of the Ancients*. Translated by H. L. Brose. Berlin, 1930.

Qian Baocong. *Suanjing Shi Shu*. Shanghai: Zhoaghua Shuju, 1963.

Rayner, W. H. "Surveying in Ancient Times." *Civil Engineering* 9 (1939): 612–14.

Reischauer, Edwin, and John K. Fairbank. *East Asia: The Great Tradition.* Volume 1 of *A History of East Asian Civilization*. New York: Houghton Mifflin, 1960.

Schöne, Hermann. *Heron Alexandrinus*. 3 vols. 1903. Stuttgart: B. G. Teubner, 1976.

Skyring-Walters, R. C. "Greek and Roman Engineering Instruments." *Transactions: The Newcomen Society for the Study of Engineering and Technology* 2 (1921): 45–60.

Swetz, Frank J. "The Amazing Chiu Chang Suan Shu." *Mathematics Teacher* 65 (1972): 425–30.

———. *Capitalism and Arithmetic: The "New Math" of the Fifteenth Century*. La Salle, Ill.: Open Court, 1989.

———. "Trigonometry Comes Out of the Shadows." In *Proceedings of the Kristiansand Workshop on the History of Mathematics, Kristiansand, Norway, August 1989*.

Swetz, Frank J., and T. I. Kao. *Was Pythagoras Chinese? An Examination of Right Triangle Theory in Ancient China*. University Park, Pa.: The Pennsylvania State University Press, 1977.

van der Waerden, B. L. *Geometry and Algebra in Ancient Civilizations.* New York: Springer-Verlag, 1983.

van Hée, L. "Le Hai-Tao Souan-King de Lieou." *Toung pao* 20 (1920): 51–60.

———. "Le Classique d l'Île Maritime: Ouvrage Chinois de IIIe siècle." *Quellen und Studien zur Geschichte der Mathematik* 2 (1932): 255–58.

Victor, S. K. *Practical Geometry in the High Middle Ages*. Philadelphia: American Philadelphia: American Philosophical Society, 1979.

Vogel, Kurt. "Ein Vermessungsproblem Reist von China nach Paris." *Historia Mathematica* 10 (1983): 360–67.

Wagner, Donald B. "Liu Hui and Tse Keng-Chih on the Volume of a Sphere." *Chinese Science* 3 (1978): 59–79.

Wu Wenchun. "Haidao Suanjing Gu Zheng Dao Yuan." In *Jiuzhang Suanshu*, ed. Wu Wenchun, pp. 162–81. Beijing: Beijing Normal University, 1982.

———. "Wo Guo Gudai Cewang Zhi Xue Chongcha Lilun Pingjji Jian Ping Shuxueshi Yanjiu Zhong Mo Xie Fangfa Wenti." *Kojishi wenji* 8 (1982): 10–30.

———. "The Out-In Complementary Principle." In Institute of the History of Natural Sciences, Chinese Academy of Science, *Ancient China's Technology and Science*, pp. 66–89. Beijing: Foreign Languages Press, 1983.

Wu Wenchun, ed. *Jiuzhang Suanshu Yu Liu Hui*. Beijing: Beijing Normal University, 1982.

Wylie, Alexander. *Notes on Chinese Literature*. Shanghai, 1867.

Zeller, M. C. *The Development of Trigonometry from Regiomontanus to Pitiscus*. Joliet, Ill.: Sisters of St. Francis of Mary Immaculate, 1946.

INDEX